CW00344016

1 MONTH OF
FREE
READING

at

www.ForgottenBooks.com

By purchasing this book you are
eligible for one month membership to
ForgottenBooks.com, giving you
unlimited access to our entire
collection of over 1,000,000 titles via
our web site and mobile apps.

To claim your free month visit:

www.forgottenbooks.com/free898183

ISBN 978-0-266-84536-2
PIBN 10898183

THE

YOUTH'S

COLUMBIAN CALCULATOR:

BEING

AN INTRODUCTORY COURSE ON

ARITHMETIC FOR BEGINNERS,

ADAPTED

TO THE CURRENCY AND PRACTICAL BUSINESS OF THE AMERICAN REPUBLIC.

FOR THE USE OF THE DISTRICT SCHOOLS.

BY ALMON TICKNOR,

AUTHOR OF THE "COLUMBIAN CALCULATOR," "ACCOUNTANT'S ASSISTANT," "MATHEMATICAL TABLES," ETC.

"Philosophy, wisdom, and liberty, support one another: he who will not reason is a *fanatic*, he who can not reason is a *fool*, he who *dare* not reason is a *slave*."

STEREOTYPE EDITION.

POTTSVILLE, PA.:

BENJAMIN BANNAN, CENTRE STREET.

PHILADELPHIA: DANIELS & SMITH.

NEW YORK: J. S. REDFIELD, CLINTON HALL.

AND FOR SALE BY ALL THE BOOKSELLERS IN THE UNITED STATES.

1848.

STEREOTYPED BY REDFIELD & SAVAGE,
13 Chambers Street, N. Y.

REMARKS.

THE first care of an author of an elementary work should be simplicity, combined with utility; but, in so doing, he should avoid the extreme, too frequently committed in some recent productions, of reducing his system *below* the standard of the *juvenile* mind, and making it *puerile*, whereby much time is required in the acquisition of *unimportant trifles*, such as require but a small portion of time, and will naturally appear perfectly plain in the regular course of instruction.

The modern system of teaching arithmetic, recently introduced by authors and teachers into our schools, called "oral," "mental," "inductive," "analytical," &c., by the division and subdivision of an *apple*, counting *balls, marbles, birds, bees*, &c., may answer a good purpose, and prove highly beneficial, in the "infant school," and, under the management of a competent instructor, prepare the way for the introduction of a systematical course of instruction by the use of the *slate.*

As an introductory course, it is always expected that the pupil will be exercised in the elementary rules of the science, sufficient to acquire some knowledge of numbers, *previous* to the use of the *slate;* but this can never exceed a mere introductory course, for no one will presume that calculations of importance can be made without the application of *rules*, and the use of the *slate.*

It is the duty of the *teacher* to resort to every *rational* and *consistent* means to communicate knowledge, and *explain* and *illustrate* his meaning, and the *theory* and *practical operation*, whether it be *oral, mental, inductive,* or *analytical.* But this *alone* is not sufficient; nor is much *time* required, for the pupil must come to the *sober realities* of the *slate* and *black-board*, otherwise it will be impossible to become master of the science—nor can we rely on the *accuracy* of *mental* calculation. It is a subject of *regret*, that so much *time* and *paper* has been employed in the introduction of this "*juvenile mathematics*" into our *arithmetics* and *schools.*

The other works on arithmetic are apparently of one family—the lineal descendants of Master Dilworth's work; the *rules, questions,* and *arrangement,* are observable in *all ;* th ▸ examples chiefly in *pounds, shillings,* and *pence*—a currency unknown to our laws ; and an unnecessary amount of *compound matter,* in place of good *practical examples*—which is a heavy tax on the *mind, time,* and *patience,* of both *teacher and pupil,* without an adequate return of *scientific instruction.*

The time employed by the pupils in the district schools in making useless calculations in pounds, shillings, and pence, would be sufficient to take them through a regular course of mathematics; for it is nothing more nor less than a reduction of *pounds* to *farthings,* and *farthings* to *pounds,* which may be acquired in *one hour.*

In France, the currency is similar to ours—namely, *decimal.* What would be the opinion of the Minister of Public Instruction, if a treatise on arithmetic should be prepared for the use of their schools, *adapted* to the *currency* of *England* or *Prussia,* to the exclusion of *their own system* ? Would the book find a place in their schools ?

This little volume has been prepared expressly for the use of *young* pupils, or those commencing the study of arithmetic; the questions are all *original,* and about eight hundred in number. It was the intention of the Author of the "Calculator," when it was first published, to prepare a work of this kind, from the fact that a work of that size is liable to be worn out before the pupil can go through it.

As that work contains a greater amount of instruction and *valuable matter* than any other of the kind extant, and embraces a full and *perfect system,* it was not particularly intended for *juvenile* scholars—at least, until they had acquired *some* knowledge of numbers.

The young student will gain more information from this *small* work than from those *large works,* now in use, whose pages are filled with anything except *suitable practical examples.* In this volume the *examples, explanations, rules, reviews,* etc., are presented in a *rational, clear, lucid,* and *pleasing* manner, such as can not but instruct. And that it may prove a benefit to those for whom it is intended, and receive the approbation of both teacher and pupil, is the sincere desire of the Author, with his best wishes for their *prosperity* and *happiness.*   A. T.

N. B. This volume, with the "Calculator," and *stereotyped* edition of the Key, will embrace about 3,500 questions for solution, nearly all of them *original.*

CONTENTS.

EXPLANATION OF SIGNS.

Signs.   *Explanations.*

= EQUAL : as, 10 mills=equal 1 cent: 10 cents=1 dime;
and when placed between two numbers, it denotes that
they are equal to each other.

+ MORE, or Addition: as, 2+2=4; it also denotes a remainder, and when placed between two or more numbers,
denotes that those numbers are to be added together; as,
2+3+4=9.

— LESS, or Subtraction; as, 9—7=2: 7 from 9 and 2 remain; when placed between two numbers, it denotes
that the number on the *right* is to be subtracted from the
number on the *left*: 10—5=5.

× INTO, or Multiplication; as, 4×4=16: 4 multiplied by
4 is 16; and when placed between two or more numbers,
denotes that they are all to be multiplied into each other:
2×2×3=12.

÷ DIVISION; as 25÷5=5: 25 divided by 5 is 5; that is,
5 is contained in 25 5 times, and 5×5=25; when placed
between two numbers, denotes that the number on the
left is to be divided by the number on the *right*:
16÷4=4.

: :: : PROPORTION: as, 2 : 4 :: 8 : 16; that is, as 2 is to 4,
so is 8 to 16; or, 16 : 8 :: 4 : 2, as 16 is to 8, so is 4 to
2, &c.

YOUTH'S COLUMBIAN CALCULATOR.

ARITHMETIC.

1. ARITHMETIC is a branch of the science of mathematics, and is the art of computing by numbers, by the operation of six rules, namely: Notation, Numeration, Addition, Subtraction, Multiplication, and Division; two of which may be considered primary rules—namely, Addition and Subtraction—and the other four secondary, as they naturally arise from the operation of the former.

2. NUMBER is that which is composed of one or more units.

NOTATION.

3. NOTATION teaches us to write or express numbers or words by the ten Arabic characters, or *digits,* so called from numbering or counting on the fingers, before the use of figures was known:—

One, two, three, four, five, six, seven, eight, nine, cipher.
1 2 3 4 5 6 7 8 9 0

4. By the use of those nine figures and ciphers, all numbers may be expressed, and their value or amount depends upon their place, or method of writing them; thus 1 is a unit, or *one:* as one apple, one book.

5. A CIPHER, when alone, is of no value; but when placed on the right of a figure or figures, it increases their value in a tenfold proportion; thus 1 and 0 are 1, but when joined (10) ten, in this manner, they become ten, which is ten *ones* (1 1 1 1 1 1 1 1 1, 10). Annex or place another cipher on the right of 10, and it is increased ten times, which is (100) one hundred, for ten tens are one hundred; then place another cipher on the right of 100, and it is increased ten times, and becomes (1000) one thousand, for ten hundred are one thousand.

6. THE VALUE of all the figures increases in the same manner: thus, 1847, the figure (7) in the place of units, denotes only its simple value (7) seven; that in the second place, or place of tens (4), is ten times its simple value, and the two figures (47) forty-seven; that in the third place, or place of hundreds (8), one hundred times its simple value (847) eight hundred and forty-seven; that in the fourth place, or place of thousands (1), one thousand times its simple value, or (1847) one thousand eight hundred and forty-seven.

REVIEW.

1. What is Arithmetic?   2. Of number?   3. Notation? 4. Of the use of figures?   5. A cipher?   6. Of the value of figures?   What is science?   What is knowledge?

There is another method of Notation by letters, viz. :—

THE ROMAN NOTATION.

I	1	one	XXIV	24	twenty-four
II	2	two	XXV	25	twenty-five
III	3	three	XXVI	26	twenty-six
IV	4	four	XXVII	27	twenty-seven
V	5	five	XXVIII	28	twenty-eight
VI	6	six	XXIX	29	twenty-nine
VII	7	seven	XXX	30	thirty
VIII	8	eight	XL	40	forty
IX	9	nine	L	50	fifty
X	10	ten	LX	60	sixty
XI	11	eleven	LXX	70	seventy
XII	12	twelve	LXXX	80	eighty
XIII	13	thirteen	XC	90	ninety
XIV	14	fourteen	C	100	one hundred
XV	15	fifteen	CC	200	two hundred
XVI	16	sixteen	CCC	300	three hundred
XVII	17	seventeen	CCCC	400	four hundred
XVIII	18	eighteen	D	500	five hundred
XIX	19	nineteen	DC	600	six hundred
XX	20	twenty	DCC	700	seven hundred
XXI	21	twenty-one	DCCC	800	eight hundred
XXII	22	twenty-two	DCCCC	900	nine hundred
XXIII	23	twenty-three	M	1000	one thousand

MDCCCXLVII   1847   one thousand eight hundred and forty-seven

Explanation.—In notation by letters, I represents one; V five; X ten; L fifty; C one hundred; D five hundred; M one thousand, &c.

As often as any letter is repeated, so many times its value is repeated, unless it be a letter representing a *less* number, placed before one representing a *greater*—then, the less number is taken from the greater: thus IV represents four; IX nine, &c., as will be seen in the above table, which the pupil should commit to memory.

NUMERATION.

1. *By Numeration*, we are taught to read any number of figures, and ascertain their relative value, when taken in connexion with each other, which is determined by the situation in which they are placed, which may be easily learned from the following tables :—

TABLE I.

hundreds of millions.	tens of millions.	millions.	hundreds of thousands.	tens of thousands.	thousands.	hundreds.	tens.	units.
9	8	7	6	5	4	3	2	1

TABLE II.

1	one.
12	twelve.
123	1 hundred and 23.
1234	1 thousand 2 hundred and 34.
12345	12 thousand 3 hundred and 45.
123456	123 thousand 4 hundred and 56.
1234567	1 million 234 thousand 5 hundred and 67.
12345678	12 millions 345 thousand 6 hundred and 78.
123456789	123 millions 456 thousand 7 hundred and 89.
1234567891	thousands of millions.
12345678912	tens of thousands of millions.
123456789123	hundreds of thousands of millions.
1234567891234	billions.

2. *To enumerate* where the numbers are large, it will be convenient to divide or separate them into periods of three figures: the first period being hundreds; the second, hundreds of thousands; the third, hundreds of millions, &c.: thus, 987,654,321. Then begin at the right, or place of *units*, and read toward the left as in table first, which is nine hundred and eighty-seven millions, six hundred and fifty-four thousand, three hundred and twenty-one.

3. When even hundreds, thousands, &c., are to be written, place ciphers on the right of 1, thus: (100), one hundred; (1,000), one thousand; (10,000), ten thousand; (100,000), one hundred thousand, &c.

Write in words the following numbers: 21, 47, 52, 86, 79, 64, 96, 109, 150, 192, 704, 879, 1001, 10004, 2468791.

Write in figures the following numbers: seventy-four; eighty-three; five hundred; six hundred and ninety-nine; seven hundred and forty-two; nine thousand, seven hundred; one million, two hundred thousand; five millions, three hundred and twenty thousand, four hundred.

REVIEW.

1. What is Numeration? Repeat the table, first beginning with units. Read Table 2, as far as the ninth line of figures. 2. How do you enumerate figures? Why do you enumerate from right to left? *Ans.* Because they increase in a tenfold proportion. How do you read figures to express their value? *Ans.* From left to right, because they decrease from left to right, in the same manner as they increase in value from right to left by enumeration. 3. How will you express or write even hundreds, thousands, &c.? Write in figures one million, two hundred thousand. Enumerate the following numbers: 10011, 70204, 600302, 5400891.

ADDITION.

1. ADDITION is the first primary rule in Arithmetic, the use of which is to ascertain the amount or sum total of two or more given numbers when put or added together.

2. *Numbers* to be added must be of the same kind or denomination. You can not add yards and dollars, but dollars can be added to dollars, and yards to yards; because 5

yards added to 2 dollars would make 7, it would express no meaning, for it is still 5 yards and 2 dollars; but 5 yards and 2 yards make 7 yards, and 5 dollars and 2 dollars make 7 dollars.

3, *The sign* used in addition is the cross (+), which means, when placed between two or more numbers, that all those numbers are to be added together; it is also called *plus*, and means more; it is sometimes used to signify a remainder, or something more.

4. *This sign* (=) signifies *equal*, or equality: as, 10 mills are=equal to 1 cent, and when placed between two numbers denotes that they are equal to each other; thus $5+5=10$, five added to five make ten, or are equal to ten; $7+6=13$; $4+3+2=9$; $7+8+5=20$; $4+5+1+3+5=18$: last example, 4 and 5 are 9 and 1 is 10 and 3 are 13, and 5 are 18.

REVIEW.

1. Which is the first primary rule in Arithmetic? What is the use of Addition? 2. When numbers are to be added, of what kind must they be? 3. What sign is used in addition? What does the sign signify besides addition? 4. What do you understand by the sign of equality? Can you give examples of the signs of addition and equality?

ADDITION TABLE.

2 and 2	are 4	3 and 9	are 12	5 and 9	are 14		8 and 8	are 16
2	3	5	3	10	13	5	10	15	8	9	17
2	4	6	3	11	14	5	11	16	8	10	18
2	5	7	3	12	15	5	12	17	8	11	19
2	6	8	4	4	8	6	6	12	8	12	20
2	7	9	4	5	9	6	7	13	9	9	18
2	8	10	4	6	10	6	8	14	9	10	19
2	9	11	4	7	11	6	9	15	9	11	20
2	10	12	4	8	12	6	10	16	9	12	21
2	11	13	4	9	13	6	11	17	10	10	20
2	12	14	4	10	14	6	12	18	10	11	21
3	3	6	4	11	15	7	7	14	10	12	22
3	4	7	4	12	16	7	8	15	11	11	22
3	5	8	5	5	10	7	9	16	11	12	23
3	6	9	5	6	11	7	10	17	12	12	24
3	7	10	5	7	12	7	11	18	13	13	26
3	8	11	5	8	13	7	12	19	14	14	28

To read the table, say 2 and 2 are 4; 3 and 3 are 6, &c. (Repeat the table.)

Exercises.

How many are 3 and 7? 4 and 5? 6 and 3? 2 and 5? 8 and 3? 2 and 8? 3 and 8? 7 and 8? 7 and 5? 4 and 10? 4 and 12; 8 and 3? 2 and 4? 4, 6, and 3? 5, 4, and 8? 7, 6, and 2? 9, 3, and 2? 7, 2, and 8? 9, 3, and 5? 11, 2, and 3? 7, 9, and 7? 8, 4, and 4? 9, 10, and 5? 10, 8, and 4? 12, 7, and 2? 11, 9, and 8? 6, 12, and 9? 11, 12, and 8? 12, 5, and 12? 5, 8, and 4? 10, 11, and 12? 11, 11, and 9?

If the teacher expects his pupils to acquire a correct knowledge of numbers, it is important that they should learn all of the *tables,* and pass a strict examination in the *reviews.* If this course is omitted, the time and labor are lost.

RULE.

(For addition of whole numbers, or integers, called Simple Addition.)

1. Set the given numbers under each other, with units under units, tens under tens, hundreds under hundreds, &c.
2. Draw a line (————) under the last number at the bottom, and begin at the right-hand column, or place of units, and add (upward) all the figures in that column, and find how many tens are contained in their sum.
3. Set down the sum when less than ten; if ten or more, set down the right-hand figure, and add the left-hand figure to the next column; if even tens, set down a cipher.
4. Proceed in this manner to the last column, and set down the whole amount of this column, which is the *sum,* or whole amount.

PROOF.

5. Perform the operation a second time, agreeably to the Rule; but in this case begin at the *top;* or, reserve one of the given numbers, find the sum of all the rest, and thereto add the number reserved. (See the 1st example.)
6. *Note.*—The reason why we carry one for every ten in whole numbers, is this; in the place of units, it requires 10 to make 1 in the place of tens; and in the place of tens, it requires 10 to make 1 in the place of hundreds: hence we always carry *one* from one denomination to another, as it requires of that denomination to make *one* in the next.

1. How will you set down numbers? 2. After the numbers are all set down, what is the next process? 3. If the column is less or more than ten? if the column is even tens, as 20, 30, &c.? 4. What is to be done with the last column? What is this last number called? *Ans.* The *sum*, or amount of all the parts, because they have all been added together and expressed in one number. 5. How will you prove the correctness of the operation? When you have proved your sum according to rule, do you know it to be correct? *Ans.* A question may be proven according to rule, and be incorrect, because if an error was committed in the first addition, the same may occur in the proof. What is the most certain method of proof? *Ans.* A correct performance of all the operation. 6. Why do you carry one for every ten?

EXAMPLES.

(1.)
```
452        452
384        384
567        567
305        305
1708 sum.  1708
           ─────
           1256
           ─────
           1708 proof.
```

Explanation.—Begin with the 5 in the place of units, and say—5 and 7 are 12, and 4 are 16, and 2 are 18; set down 8 and carry 1 to 0 is 1, and 6 are 7, and 8 are 15, and 5 are 20; this is even tens: set down 0 and carry 2 to 3 are 5, and 5 are 10, and 3 are 13, and 4 are 17; set down the whole of the last column; and all of the numbers added amount to 1708, which is the sum or answer. Then to prove it, draw a line under this sum, and another under the number at the top, which number omit in the addition, and add only the three numbers, thus: 5 and 7 are 12 and 4 are 16; set down 6, and carry 1 to 0 is 1, and 6 are 7, and 8 are 15; set down 5, and carry 1 to 3 is 4, and 5 are 9, and 3 are 12, which set down=1256: add this last number and the number at the top together, 6 and 2 are 8, 5 and 5 are 10, set down 0, and carry 1 to 2 are 3, and 4 are 7, and 1 is 1,= 17,=1708. It is not necessary to write the sum twice— once is sufficient.

(2)

(2.) 45	(3.) 78	(4.) 580	(5.) 437	(6.) 504	(7.) 8042
32	43	724	648	358	3587
50	21	685	750	415	6435
49	53	712	321	324	4789
35	70	304	405	502	6504
67	41	531	123	423	2130

(8.)	(9.)	(10.)	(11.)	(12.)	(13.)
421	689	804	7098	7408	6503
384	742	795	4625	4976	4764
796	381	642	8370	8532	5041
342	709	358	5064	4576	3258
481	476	764	8521	5803	7890
798	879	207	3450	4765	5476

(14.) cents.	(15.) dollars.	(16.) seconds.	(17.) minutes.	(18.) hours.	(19.) days.
4786	316	789	7404	3004	4789
2174	704	684	8241	7021	7468
3155	589	879	5720	8796	6478
7809	876	658	7804	5845	3040
7456	942	847	5796	6782	5060
2104	758	602	4235	3070	7047

(20.) 289+786+479+642+751.   *Ans.* 2947.
(21.) 334+798+207+658+207.   2204.
(22.) 704+402+350+764+208.   2428.
(23.) 275+680+542+856+764.   3117.
(24.) 507+647+372+214+656.   2396.

(25.)	(26.)	(27.)	(28.)	(29.)
21478	41034	50876	60478	31456
65420	79965	43215	3214	78504
34042	42071	87058	305	32150
58376	32458	65421	78	321
21054	76542	32507	4	4780
67892	37056	61472	8176	51347

(30.) What is the sum of 678, 947, 652, 870? *Ans.* 2156.
(31.) What is the sum of 704, 685, 770, 319 ? 2478.
(32.) What is the sum of 648, 309, 214, 762 ? 1933.
(33.) What is the sum of 218, 917, 653, 247 ? 2035.
(34.) What is the sum of 414, 702, 315, 876 ? 307.
(35.) In 5 canal-boats, the 1st is loaded with 27649 pounds; the 2d with 31974 pounds; the 3d with 34640 pounds; the 4th with 33071 pounds; the 5th with 30720 pounds. Required the number of pounds in the 5 boats. *Ans.* 157154 pounds.

(36.) The collector in this place received of A. 421 dollars; of B. 670 dollars; of C. 429 dollars; of D. 50 dollars; of E. 540 dollars; of F. 601 dollars. Required the whole amount ? *Ans.* 2711 dollars.

(37.) There are 60 seconds in one minute; how many seconds are there in 5 minutes ? *Ans.* 300 seconds.

(38.) A country merchant purchased in New York 2741 pounds of coffee, 478 pounds of loaf-sugar, 47 pounds of black tea, 321 pounds of green tea, 250 pounds of iron, 229 pounds of cut-nails, 640 pounds of salt; required the number of pounds. *Ans.* 4706 pounds.

(39.) One half of a farm is worth 15678 dollars; how much is the whole farm worth ? *Ans.* 31356 dollars.

(40.) If you should commence a journey and travel 7 days, the first day 32 miles, and then increase 1 mile every day, how many miles would you travel in the 7 days ? *Ans.* 245 miles.

(41.) A gentleman near Cincinnati (Ohio), who owns an extensive plantation, raised in one season 4780 bushels of wheat, 2070 bushels of rye, 7892 bushels of corn, 3879 bushels of oats, 1500 bushels of turnips, 2798 bushels of potatoes, 640 bushels of buckwheat, 5 bushels of cloverseed, and 7 bushels of flaxseed; how many bushels did the plantation produce in one year ? *Ans.* 23571 bushels.

(42.) Add together 6079, 30, 4789, 607, 52, 8496. *Ans.* 20053.

(43.) Add together 31, 47, 608, 9742, 39, 78965. *Ans.* 89432.

(44.) Add together 7, 92, 87, 640, 5617, 87954. *Ans.* 94397.

(45.) Add together 96145, 3045, 281, 79, 84, 9. *Ans.* 99843.

(46.) A. has six hundred and forty-eight sheep; B. has nine hundred and eighty-seven; C. has fourteen hundred and ninety-one; D. has two thousand: how many sheep in all four of the flocks?   *Ans.* 5126 sheep.

(47.) A merchant purchased of D. 642 barrels of flour, for which he paid 4792 dollars; of G. 783 barrels, for which he paid 5380 dollars: how many barrels of flour did he purchase, and how much did he pay?

     Ans. 1425 barrels, and paid 10172 dollars.

(48.)	(49.)	(50.)	(51.)	(52)
980467	845067	380976	748650	689075
351248	456796	372045	429767	697236
375896	217904	487650	321586	480576
456970	321580	487321	479605	217856
834276	654234	548653	803072	232051
854987	215087	214876	542143	678970

SUBTRACTION.

1, SUBTRACTION is the second primary rule in Arithmetic, and is the reverse of Addition. 2. It teaches to take a less number from a greater of the same name or kind, and to show their difference, or remainder.

3. There must always be two given numbers in subtraction.

4. The larger number is called the *minuend,* and the less the *subtrahend.* 5. The difference between those two numbers is called the *remainder.*

6. The sign (—) which is called *minus* or *less,* when placed between two numbers, signifies that the number on the right is to be subtracted from the number on the left.

7. Thus 10—5=5 remainder; that is, 5 subtracted from 10 will leave 5, and this is the difference, or remainder.

8. If you have 10 dollars, and should pay a debt of 5 dollars, you would have 5 remaining, because 5+5=10.

REVIEW.

1. Which is the second primary rule? What is it the reverse of? 2. What does it teach? 3. How many numbers must be given? 4. What are they called? 5. What is the difference called? 6. What is the sign, and what does it signify? 7. Explain this operation. 8. If you take 5 from 10, why would 5 remain?

SUBTRACTION TABLE.

1—1=0		3—3=0		5—5=0		7—7=0		9—9=0	
2	1	4	1	6	1	8	1	10	1
3	2	5	2	7	2	9	2	11	2
4	3	6	3	8	3	10	3	12	3
5	4	7	4	9	4	11	4	13	4
6	5	8	5	10	5	12	5	14	5
7	6	9	6	11	6	13	6	15	6
8	7	10	7	12	7	14	7	16	7
9	8	11	8	13	8	15	8	17	8
10	9	12	9	14	9	16	9	18	9
11	10	13	10	15	10	17	10	19	10
12	11	14	11	16	11	18	11	20	11
13	12	15	12	17	12	19	12	21	12
14	13	16	13	18	13	20	13	22	13
15	14	17	14	19	14	21	14	23	14
16	15	18	15	20	15	22	15	24	15
17	16	19	16	21	16	23	16	25	16
18	17	20	17	22	17	24	17	26	17
19	18	21	18	23	18	25	18	27	18
20	19	22	19	24	19	26	19	28	19

2—2=0		4—4=0		6—6=0		8—8=0		10—10=0	
3	1	5	1	7	1	9	1	11	1
4	2	6	2	8	2	10	2	12	2
5	3	7	3	9	3	11	3	13	3
6	4	8	4	10	4	12	4	14	4
7	5	9	5	11	5	13	5	15	5
8	6	10	6	12	6	14	6	16	6
9	7	11	7	13	7	15	7	17	7
10	8	12	8	14	8	16	8	18	8
11	9	13	9	15	9	17	9	19	9
12	10	14	10	16	10	18	10	20	10
13	11	15	11	17	11	19	11	21	11
14	12	16	12	18	12	20	12	22	12
15	13	17	13	19	13	21	13	23	13
16	14	18	14	20	14	22	14	24	14
17	15	19	15	21	15	23	15	25	15
18	16	20	16	22	16	24	16	26	16
19	17	21	17	23	17	25	17	27	17
20	18	22	18	24	18	26	18	28	18

To read the table, say 1 from 2 and 1 remains; 1 from 5 and 4 remain.

(2*)

Exercises.

If 3 be taken from 8, how many will be left? 2 from 7?
7 from 9? 5 from 9? 4 from 9? 2 from 9? 6 from 9? 8 from
9? 2 from 8? 4 from 8? 3 from 8? 6 from 8? 7 from 8? 1
from 7? 4 from 7? 3 from 7? 5 from 7? 7 from 10? 9 from
11? 7 from 12? 5 from 12? 4 from 11? 6 from 12? 7 from
13? 8 from 14? 9 from 13? 10 from 14? 9 from 15? 11
from 16? 8 from 16? 9 from 15? 9 from 20? 11 from 20?
12 from 18? 11 from 19? 13 from 15? 11 from 17? 12
from 15? 11 from 13? 7 from 19?

Take 7 from 12 and 2 from the remainder.
Take 5 from 14 and 5 from the remainder.
Take 7 from 16 and 4 from the remainder.
Take 8 from 12 and 4 from the remainder.

RULE I.

1. Write down the greatest number first, then write the
less number directly under it, observing to place units under units, tens under tens, &c.; draw a line underneath.

2. Begin with the units, or right-hand figure, and subtract
that figure from the figure over it, and set down the difference.

3. When the figure in the lower number is more than the
one above it, subtract from 10, and the difference between
that figure and 10 must be added to the figure in the upper
number; then set down that figure.

4. When you subtract from 10, carry 1, and add it to the
next left-hand figure. 5. Proceed in this manner with all
the figures, and the number thus obtained will be the difference between the two given numbers.

RULE II.

1. After stating the sum as above directed, then, if either
of the lower figures be greater than the upper one, conceive
10 to be added, or add 10 to the upper figure; then take the
lower figure from it, and set down the remainder.

2. When 10 is thus added to the upper figure, there must
be 1 added to the next lower figure.

PROOF.

Add the remainder, or difference, to the less number, and
their sum will be equal to the greater number.

1. How will you write numbers in Subtraction? 2. Where do you begin to subtract? 3. When the figure in the lower number is more than the one above it, how will you proceed? 4. When you subtract from 10, or *borrow*, what must be done? 5. What next? RULE 2: 1. How can you subtract by this rule? 2. When 10 is added to the upper figure, what must be done? Proof.

Take 13 from 24, and 3 from the remainder.
Take 15 from 27, and 4 from the remainder.
Take 17 from 30, and 5 from the remainder.
Take 20 from 29, and 3 from the remainder.

Examples.

(1.) 8407 minuend.
7325 subtrahend.
1082 remainder.
8407 proof.

Explanation.—Begin in the place of units with 5, and say 5 from 7 and 2 remain, which set down; then 2 from 0 you can not, but 2 from 10 and 8 will remain, set this down; now there is 1 to carry to the next figure on the left, 3, which added to it will make 4; then 4 from 4 and 0 remain; then 7 from 8 and 1 remain. This last number, 1082, is the difference in value between the two numbers above the line, *minuend* and *subtrahend:* because if we add this remainder, or difference, to the *subtrahend*, the number subtracted from the minuend, it will produce that number; therefore, draw a line under the remainder and add it to the subtrahend. Thus 2 and 5 are 7, 8 and 2 are 10, set down 0, and carry 1 to 0 is 1, and 3 are 4, 1 and 7 are 8=8407 proof: the same as the minuend.

	(2.)	(3.)	(4.)	(5.)	(6.)
From	789706	96845	68409	164089	648978
Take	548612	62407	42387	32147	209546
Rem.	241094	34438	26022	131942	439432
Proof.	789706	96845	68409	164089	648978

(7.)	(8.)	(9.)	(10.)	(11.)
940876	804256	704523	84765	65389
207904	249087	215046	2804	40856

(12.) 9870796 — 8405962 rem.1464834	(27.) 9789658 — 4981	
(13.) 6058962 — 5039780 1019182	(28.) 4259076 — 6009	
(14.) 3179845 — 2349007 830838	(29.) 4158430 — 7409	
(15.) 8119764 — 7209840 909924	(30.) 6569421 — 60091	
(16.) 6008971 — 5006472 1002499	(31.) 598040 — 3989	
(17.) 8429060 — 7984597 444463	(32.) 7009100 — 79945	
(18.) 7980065 — 6400458 1579607	(33.) 6580910 — 64208	
(19.) 8402510 — 7098079 1304431	(34.) 5670959 — 30098	
(20.) 4807610 — 3907009 900601	(35.) 6789765 — 999	
(21.) 3450789 — 1458062 1992727	(36.) 4321459 — 7890	
(22.) 6479870 — 5476801 1003069	(37.) 6487071 — 8979	
(23.) 1230245 — 1198760 31485	(38.) 5214861 — 8458	
(24.) 5830796 — 4804976 1025820	(39.) 3145670 — 68798	
(25.) 4201451 — 3098207 1103244	(40.) 7490910 — 40094	
(26.) 3698079 — 2478090 1219989	(41.) 8456780 — 7589	

(42.) If you have 2190 bushels of wheat, and 907 bushels of corn, how many more bushels of wheat have you than corn?   *Ans.* 1283.

(43.) If 54798 bricks are required to build a house, and you have received 29094, how many more are required?
Ans. 25704.

(44.) General Washington was born in 1732; how many years from that period to 1847?   *Ans.* 115 years.

(45.) How many years since the discovery of America by Columbus, in 1492, to the present time (1847)?
Ans. 355 years.

(46.) The population of Maryland is 470019; Virginia, 1239797: how many more has Virginia than Maryland?
Ans. 769778.

(47.) A. has 7965 dollars—he will pay 3790; B. has 8742 dollars—he will pay 4071: which will have the most money after the payments?

Ans. B. will have 496 dollars more than A.

(48.) A. sold two farms for 7492 dollars each; and B. sold one farm for 14984 dollars: which will receive the most money?

(49.) A merchant purchased goods to the amount of 42692 dollars, but, being damaged, he must lose 1597 dollars on them; for how much must he sell them?

Ans. 41095 dollars.

(50.) What is the difference between 7987645 and 47815?

(51.) A grocer purchased 127 hogsheads of sugar weighing 106840 pounds, and sold 43 hogsheads weighing 48620 pounds; how many hogsheads has he left, and what is their weight?

Ans. 84 hogsheads; and their weight is 58220 pounds.

(52.) The larger of two numbers is 6497, and their difference is 4281; what is the less number? *Ans.* 2216.

(53.) How much must be added to 947 to make 1691?

Ans. 744.

(54.) A man borrowed 1200 dollars and paid 742; how many dollars did he then owe? *Ans.* 458.

(55.) A man has an estate of 25642; if he should give 3640 dollars to each of his two daughters, and the remainder to an only son, how much would the son receive?

Ans. 18362 dollars.

(56.) What is the difference between 1000001 and 10009?

Ans. 989992.

(57.) From 7070707070 take 909090909.

ADDITION AND SUBTRACTION.

(58.) Add 97, 804, 607, 3414, 7111; then subtract 8409.

Ans. 3624.

(59.) Add 6092+7084+49+68+400+97840−50946.

Ans. 60587.

(60.) A clerk went out to collect money: he collected of one man 150 dollars; of another, 124 dollars; of another, 221 dollars; and paid a debt, out of the money collected, of 97 dollars: how much did he have left? *Ans.* 398 dollars.

(61.) A gentleman wishing to establish his son in business, had deposited in bank 5692 dollars, for which he drew checks as follows: one of 984 dollars, one of 1296 dollars, one of 1421 dollars, and one of 789 dollars; after receiving those four sums, how much would remain in bank?

Ans. 1202 dollars.

(62.) Mr. Brown holds a note-of-hand against Mr. White for 2692 dollars; and Mr. White has paid, at one time 794 dollars, at another 681 dollars, at another 459 dollars: now, how much does Mr. White owe Mr. Brown on the note?

Ans. 758 dollars.

(63.) From the sum of 6804978, 794604, subtract the sum of 408, 976, 432, 978, 6458, 79642. *Ans.* 7510688.

(64.) From the sum of 40987642, 789765, subtract the sum of 78094, 760298. *Ans.* 40939015.

(65.) Maine has 501793 inhabitants; New Hampshire, 284574; Vermont, 291948; Massachusetts, 737699: how

many more inhabitants have Vermont and Massachusetts
than Maine and New Hampshire?   *Ans.* 243280.

(66.) The population of the city of New York is 331642,
and that of Philadelphia 263741; required the difference.
Ans. 67901.

(67.) If you add 760, 5840, 72, 96, 412V, and then subtract
that amount from 59784, how much will remain?
Ans. 48889.

(68.) If you should lend your friend 3 sums of money—
the first, 500 dollars; the second, 642; the third, 489—and
he should make two payments, one of 398 dollars, the second
999 dollars, how much does your friend owe you?
Ans. 234 dollars.

(69.) A farmer has 3 farms; the 1st contains 640 acres,
the 2d 221 acres, the 3d 384 acres; the 1st is worth 18492
dollars, the 2d 7849 dollars, the 3d 11820 dollars: how
many acres has he in all, and how much are they worth? if
he should sell 629 acres, how many would he have left?
Ans. Number of acres, 1245−629=616 acres left;
worth of the 3 farms, 38161 dollars.

(70.) 78964215+642790420−4567842+4850976.

MULTIPLICATION.

1. By MULTIPLICATION, we can perform a number of additions by a shorter and more easy method; or it is a number repeated a given number of times: for if we multiply 5
by 5, the product is 25, because the 5 is repeated 5 times;
so if we add five 5s together, it would be the same, thus:
5+5+5+5+5=25.

2. There are three parts in Multiplication, which require
particular attention:—

1. The *multiplicand*, the number given to be multiplied.

2. The *multiplier*, the (less) number by which you multiply.

3. The *product*, which is the result, or sum produced by
the operation of multiplying, and is just as many times
larger than the *multiplicand* as there are *units* in the multiplier.

4. The multiplicand and multiplier, *together*, are called
factors.

MUL[TIPLICATION] TABLE.

Twice		3 times		[4] times		5 times	
1 make 2		1 [m]ake [3]		1 make 4		1 make 5	
2	4	2		2	8	2	10
3	6	3		3	1[2]	3	15
4	8	4		4		4	20
5	10		5	5	20	5	25
6	12		18	6	24	6	30
7	14	7	21	7	28	7	35
8	16	8	24	8	32	8	40
9	18	9	27	9	36	9	45
10	20	10			40	10	50
11	22	11		11	44	11	55
12	24	12		12	48	12	60

6 times		7 times		8 times		9 times	
1 make 6		1 make [7]		1 make 8		1 make 9	
2	12	2	14	2	16	2	18
3	18	3	21	3	24	3	27
4	24	4	28	4	32	4	36
5	30	5	35	5	40	5	45
6	36	6	42	6	48	6	54
7	42	7	49	7	56	7	63
8	48	8	56	8	64	8	72
9	54	9	63	9	72	9	81
10	60	10	70	10	80	10	90
11	66	11	77	11	88	11	99
12	72	12	84	12	96	12	108

10 times		11 times		12 times	
1 make 10		1 make 11		1 make 12	
2	20	2	22	2	24
3	30	3	33	3	36
4	40	4	44	4	48
5	50	5	55	5	60
6	60	6	66	6	72
7	70	7	77	7	84
8	80	8	88	8	96
9	90	9	99	9	108
10	100	10	110	10	120
11	110	11	121	11	132
12	120	12	132	12	144

REVIEW.

1. What is Multiplication? Which is the most convenient—to multiply several numbers, or to add them together?
2. How many parts are there in Multiplication? Name them. 1. What is the multiplicand? 2. What is the multiplier? 3. The product? How many times larger is the product than the multiplicand? 4. By what name are multiplier and muliplicand together called? How many terms, or numbers, are given in multiplication? *Ans.* Two—mul- tiplicand and multiplier. How many are required? *Ans.* Three; by multiplying the two given numbers together, the third is produced, which is the *product*, called also the *answer*, to the question. Repeat the table, beginning with 1, to 144, then begin with 144 and repeat it *back :* the reason of this is evident—it will enable you to multiply with ease and facility; then there will be no occasion for counting *fingers*, nor straight *marks* on the *slate*, which is a very improper practice.

Exercises.

In Multiplication, we use this sign (×), and when written between two or more numbers, it means that they are to be multiplied together; thus $3 \times 3 = 9$; $3 \times 3 \times 3 = 27$; $3 \times 3 = 9 \times 3 = 27$, &c.

Multiply 2 by 4; 4 by 3; 5 by 3; 3 by 6; 4 by 7; 9 by 4; 8 by 5; 6 by 8; 7 by 9; 6 by 9; 9 by 9; 9 by 8; 10 by 11; 11 by 9; 10 by 4; 11 by 7; 11 by 11; 6 by 11; 5 by 2, by 3; 4 by 6, by 4; 7 by 2, by 6; 4 by 8, by 2; 7 by 2, by 3; 5 by 4, by 6; 8 by 3, by 2; 11 by 4, by 3; 5 by 2, by 5; $4 \times 2 \times 8$; $3 \times 2 \times 4$; $7 \times 2 \times 5$; $4 \times 2 \times 3$; $7 \times 4 \times 3$; $5 \times 4 \times 3 \times 2$.

RULE I.

(When the multiplier does not exceed 12.)

1. Write down the multiplicand, and at the right, beginning with units and under the figures of the multiplicand, write the multiplier.

2. Then begin with the units, and multiply all the figures in the multiplicand in succession, and set down their several products, observing to carry *one* for every *ten* to the product of the next figure, and set down the whole of the last product.

Proof. Multiply the multiplier by the multiplicand; however, the most correct method of proof is by Division, as soon as the pupil has learned Division.

EXAMPLES.

(1.) 7405   multiplicand.   *Explanation.*—Begin at the
     3 ×   multiplier.   right-hand with 3, and say 3
   22215   product.   times 5 are 15, set down 5;
                         now 3 times 0 is 0, but 1 to carry
from the place of units is 1; then 3 times 4 are 12, set down
the 2; then 3 times 7 are 21, and 1 to carry from the last
product is 22, set down the whole amount.

Now the multiplicand, 7405, has been repeated 3 times,
and produced 22215, which is just 3 times as much as 7405,
and 3 is contained in 22215 just 7405 times.

(2.)	(3.)	(4.)	(5.)	(6.)
146876	2140567	5678420	6071234	405678
2 ×	3 ×	4 ×	5 ×	6 ×
293752	6421701	22713680	30356170	2434068

(7.)	(8.)	(9.)	(10.)	(11.)
890765	6489765	6849730	7085678	70890640
7 ×	8 ×	9 ×	10 ×	11 ×

(12.)	(13.)	(14.)	(15.)	(16.)
7604895	9074658	658798	1760459	4589780
12 ×	9 ×	7 ×	8 ×	12 ×

(17.)   6845756 × 9 =   61611804
(18.)   50487654 × 7 =   353413578
(19.)   670912358 × 4 = 2683649432
(20.)   645879352 × 6 = 3875276112
(21.)   308457654 × 8 = 2467661232
(22.)   798765803 × 7 = 5591360621
(23.) 1415161718 × 5 = 7075808590
(24.) 3087645401 × 3 = 9262936203
(25.) 7189700560 × 4 = 28758802240
(26.)   47658421 × 3 =   142975263

(3)

(27.)	847658934567 × 4	(33.)	748967487654 × 12
(28.)	6543021500384 × 5	(34.)	6508976583581 × 7
(29.)	74056542008971 × 6	(35.)	145100789671 × 8
(30.)	65711007456780 × 9	(36.)	84976548760 × 11
(31.)	21104956789870 × 10	(37.)	64058798765 × 12
(32.)	4687908545861 × 11		

(38.) Multiply 6742 by 2 and by 4; 6742×2=13484; 6742×4=26968; then add 13484+26968=   *Ans.* 40452.

(39.) Multiply 47023 by 3 and by 2.   235115.

(40.) Multiply 31471 by 3 and by 4.   220297.

(41.) Multiply 50624 by 4 and by 5.   455616.

(42.) Multiply 31421 by 5 and by 6.   345631.

RULE II.

(When the multiplier exceeds 12, and consists of two or more figures.)

1. Write the multiplicand, and under it the multiplier, so that units may stand under units, tens under tens, hundreds under hundreds, &c., then draw a line under them.

2. Multiply each figure in the multiplicand by each figure in the multiplier separately, beginning at the right, or place of units, placing the result directly under the multiplier, observing to carry 1 for every 10, &c.

3. Then multiply by the next figure in the place of *tens,* placing the *first* figure of every line directly under its respective multiplier.

4. After multiplying by all the figures, add these products together, and their sum, or amount, will be the product or answer required.

5. When ciphers occur at the right-hand of either the multiplicand or multiplier, or both, omit them in the operation, and *annex* them to the product. *(Annex,* to subjoin at the end, right-hand.)

6. When there are ciphers between the significant figures of the multiplier (4107), we may omit the ciphers, multiplying by the *significant figures only,* placing the first figure of each product directly under the figure by which you multiply.

REVIEW.

Rule 1.—1. How do you write down numbers in Multiplication? 2. After you have written the multiplicand and multiplier, where will you begin to multiply? Why do you carry one for every ten? Proof. What other method of proof? *Rule 2.*—1. How will you write the terms in this rule? 2. How will you multiply? 3. How will you multiply in the place of tens? 4. After you have multiplied by all the figures? 5. When ciphers occur at the right-hand? What do you understand by *annex*? 6. When there are ciphers between the significant figures, what will be done? What do you understand by *significant figures*? *Ans.* If 405 is the multiplier, I will first multiply by 5, and then pass to the 4, both of which are *significant* figures, and write the result directly under each.

EXAMPLES.

(43.)	(44.)	(45.)	(46.)	(47.)	(48.)	(49.)
15	16	18	70	60	100	45
16 ×	15	14	40	21	62	25
90	80	72	2800	60	200	225
15	16	18		120	600	90
240	240	252		1260	6200	1125

(50.)	(51.)	(52)	(53.)	(54.)	(55.)
68	304	645	741	572	320
24 ×	25	102	38	42	80
272	1520	1290	5928	1144	25600
136	608	645	2223	2288	
1632	7600	65790	28158	24024	

Remarks.—In the 52 examples, the pupil may believe that the 0 in the multiplier is of no use, as it is not repeated in multiplying; but he will observe that the product of the next figure is removed one place farther to the left, which increases the product in a ten-fold proportion; if the 0 was omitted, the multiplier would be but 12, and the product but 7740. In example 46, you could say 4 × 7 = 28, then *annex* two (00) = 2800; in example 48, you could *annex* two (00) to 62, thus 6200, and the work is done.

(56.)	(57.)	(58.)	(59.)	(60)	(61.)
14876	64856	70148	21845	45678	478914
25 ×	32	45	62	71	82
74380	129712	350740	43690	45678	957828
29752	194568	280592	131170	319746	3831312

(62.)	684892 ×	84 =	57530928	(72.) 8697845 ×	304
(63.)	500478 ×	47 =	23522466	(73.) 6589786 ×	4005
(64.)	768498 ×	29 =	22286442	(74.) 7900864 ×	701
(65.)	897860 ×	72 =	64645920	(75.) 6508970 ×	847
(66.)	842198 ×	111 =	93483978	(76.) 58079112 ×	905
(67.)	470095 ×	112 =	52650640	(77.) 3184678 ×	894
(68.)	301147 ×	122 =	36739934	(78.) 5047842 ×	1400
(69.)	478905 ×	152 =	72793560	(79.) 6584567 ×	1242
(70.)	9845601 ×	422 =	415484362	(80.) 6897050 ×	741
(71.)	7580967 ×	223 =	1690555641	(81.) 3148784 ×	6143

(When the multiplier is exactly equal to the product of any two figures in the multiplication table.)

RULE.

Multiply first by one of those figures, and that product by the other, and the last product will be the answer.

Thus, 251×16: $(4 \times 4 = 16)$ $251 \times 4 = 1004 \times 4 = 4016$; $251 \times 16 = 4016$.

(82.)	(83.)	(84.)	(85.)	(86.)
451 × 63	640 × 49	423 × 25	742 × 81	621 × 64
7	7	5	9	8
3157	4480	2115	6678	4968
9	7	5	9	8
28413	31360	10575	61102	39744

(87.)	642134 ×	56 =	(7 × 8)	*Ans.* 35959504.
(88.)	530872 ×	72 =	(8 × 9)	38222784.
(89.)	481564 ×	144 =	(12 × 12)	69345216.
(90.)	314678 ×	36 =	(6 × 6)	11328408.
(91.)	478045 ×	110 =	(10 × 11)	52584950.

To multiply by 10, 100, 1000, &c., annex all the ciphers to the multiplicand that belong to the multiplier, and it will be the product required.

(92.)	671 ×	10 =	*Ans.* 6710.
(93.)	7432 ×	100 =	743200.
(94.)	8749 ×	1000 =	8749000.
(95.)	19861 ×	10000 =	198610000.
(96.)	25000 ×	100000 =	2500000000.

To multiply by any number of nines in one line, as—9, 99, 999, &c.)

RULE.

Annex as many ciphers to the multiplicand as there are nes in the multiplier, and from this number subtract the ultiplicand; the difference will be the product required.

(97.) Multiply 78647 by 99, thus :   7864700 : annex 00.
(98.) Multiply 6849784 by 999.   78647
(99.) Multiply 31478964 by 9999.   7786053 product.

(100.) Samuel was employed by the day, at 27 cents -day, for 92 days; how many cents did he earn ?
Ans. 2484 cents.

(101.) If 32 quarts make 1 bushel, how many quarts are here in 25 bushels of wheat ?   *Ans.* 800 quarts.

(102.) If 64 pints make a bushel, required the number f pints in 18 bushels ?   *Ans.* 1152 pints.

(103.) If 112 pounds make 1 hundred weight (gross), ow many pounds would there be in 12 hundred weight ?
Ans. 1344 pounds.

(104.) If 16 ounces make 1 pound, how many ounces in 451 pounds ?   *Ans.* 23216 ounces.

(105.) If 16 drams make 1 ounce, how many drams in 1297 ounces ?   *Ans.* 20752 ounces.

(106.) If 4 quarters make a yard, how many quarters in 14984 yards ?   *Ans.* 59936 quarters.

(107.) If 4 nails make one quarter of a yard, how many ails in 16985 quarters ?   *Ans.* 67940 nails.

(108.) If 365 days make 1 year, how many days in 225 ears ?

(109.) If 12 months make a year, how many months in 145 years ?

(110.) If 7 days make a week, how many days in 167 veeks ?

(111.) If 24 hours make a day, how many hours in 674 days ?

(112.) If 60 minutes make an hour, how many minutes in 250 hours ?

(3*)

(113.) If 60 seconds make a minute, how many seconds in 444 minutes ?

(114.) If 12 inches make a foot, how many inches in 678 feet ?

(115.) If 3 feet make a yard, how many feet in 8976 yards ?

(116.) If 40 poles make a furlong, how many poles in 347 furlongs ?

(117.) If 10 dollars make an eagle, how many dollars in 121 eagles ?   *Ans.* 1210.

(118.) If 10 dimes make a dollar, how many dimes in 147 dollars ?   *Ans.* 1470.

(119.) If 10 cents make a dime, how many cents in 254 dimes ?   *Ans.* 2540.

(120.) If 10 mills make a cent, how many mills in 345 cents ?   *Ans.* 3450.

(121.) If 320 rods make a mile, how many rods in 492 miles ?   *Ans.* 157440.

(122.) If 160 square rods make an acre, how many rods in 125 acres ?   *Ans.* 20000.

(123.) If 128 feet make a cord of wood, how many feet in 68 cords ?   *Ans.* 8704.

(124.) If 1 dozen of buttons cost 23 cents, how much will 34 dozen cost ?

(125.) If 1 pound of butter cost 19 cents, how much will 120 pounds cost ?   *Ans.* 2280 cents.

(126.) If a man should travel 164 days, at the rate of 35 miles a-day, how many miles would he travel in that time ?   *Ans.* 5740.

(127.) If a company of 72 men should each receive 54 dollars as their wages, how much would they all receive ?   *Ans.* 3888 dollars.

(128.) What sum of money must be divided among 447 men, so that each man may have 25 dollars ?   *Ans.* 11175 dollars.

(129.) There are 16 houses, each house has 28 windows, and each window 12 panes of glass; how many panes of glass in the 16 houses ?   -   *Ans.* 5376.

ADDITION AND MULTIPLICATION.

(130.) Multiply 21 by 14; 19 by 31; 25 by 35: and add their products.   *Ans.* 1758.

(131.) A drover purchased 57 oxen for 62 dollars each,

and 120 cows for 24 dollars each; how many cattle did he purchase, and how much did they cost?
Ans. 177 cattle; cost 6414 dollars.

(132.) A merchant purchased 14 pieces of calico; 8 pieces contained 24 yards each, and 6 pieces 32 yards each: how many yards in all? Ans. 384 yards.

(133.) Multiply 150 by 62; 78×41; 27×31: and add their products. Ans. 13335.

(134.) How much must you add to 6495 to make it 96452? Ans. 89957.

SUBTRACTION AND MULTIPLICATION.

(135.) Multiply 387 by 640; from the product subtract 7846. Ans. 239834.

(136.) A miller purchased 12 bags of wheat, each bag weighed 245 pounds, and the bags without the wheat weighed 76 pounds; required the weight of the wheat without the bags. Ans. 2864 pounds.

(137.) Multiply 764 by 250; from the product subtract 6894. Ans. 184106.

SHORT DIVISION.

1. Division teaches to divide a larger number, by a less, into equal parts, and is a short method of performing a number of subtractions.

2. If we desire to divide 25 dollars among 5 men equally, instead of subtracting 5 from 25 five times, we could divide 25 by 5, which would give 5 dollars to each man; because 5 is contained in 25 just 5 times, and 5×5 is = 25.

3. There are 4 parts or terms in Division, that require attention, namely: Dividend, Divisor, Quotient, and Remainder.

1. *Dividend.* The number given to be divided, which is always more than the divisor.

2. *Divisor.* The given number by which the dividend is to be divided.

3. *Quotient.* That which gives the number of times the divisor is contained in the dividend, which is also the *answer*.

· **4. _Remainder_.** That which remains after all the figures
have been brought down from the dividend and divided, and
the last subtraction performed (if any remain) will be of the
same denomination with the dividend, and always _less_ than
the divisor.

REVIEW.

1. What is Division? 2. How do you know that 5 dollars is the just share of each man? 3. How many parts
in Division? 1. Explain the Dividend. 2. Which is the
Divisor? 3. What of the Quotient? 4. What of the
Remainder?

Exercises.

Divide 6 by 3; 8 by 4; 10 by 5; 12 by 6; 14 by 7; 16
by 8; 4 by 2; 12 by 3; 12 by 4; 12 by 2; 10 by 3; 9 by
5; 12 by 8; 10 by 4; 11 by 5.

RULE.

(When the divisor does not exceed 12.)

1. Write down the dividend, and draw a curved line at
the left-hand side, and a straight line under the dividend,
and place the divisor at the left-hand of it, thus:—

Divisor, 9)81   dividend.

9   quotient.

2. Consider how many times the divisor is contained in
the first figure or figures of the dividend, and set down the
result, observing whether there be any remainder, and, if
any, carry it to the left of the next figure, and consider it
placed there as so many tens, into which divide as before,
&c. 3. But, if no remainder, see how many times the divisor is contained in the next figure; if the divisor is not contained in the next figure of the dividend, write a cipher in
the quotient, and take another figure.

Proof.—Multiply the quotient by the divisor, and add in
the remainder, if any, and the product will agree with the
dividend, if the operation has been correctly performed.

4. This sign ÷ is used in division, and when placed between two numbers (9÷3), means that the number on the
left shall be divided by the one on the right, thus:—

3)9

3 times.

DIVISION TABLE.

2 in 2	1	0 rem.		4 in 5	1	1		6 in 12	2	0
2	3	1	1	4	6	1	2	7	7	1	0
2	4	2	0	4	7	1	3	7	8	1	1
2	5	2	1	4	8	2	0	7	9	1	2
2	6	3	0	4	9	2	1	7	10	1	3
2	7	3	1	4	10	2	2	7	11	1	4
2	8	4	0	4	11	2	3	7	12	1	5
2	9	4	1	4	12	3	0	8	8	1	0
2	10	5	0	5	5	1	0	8	9	1	1
2	11	5	1	5	6	1	1	8	10	1	2
2	12	6	0	5	7	1	2	8	11	1	3
3	3	1	0	5	8	1	3	8	12	1	4
3	4	1	1	5	9	1	4	9	9	1	0
3	5	1	2	5	10	2	0	9	10	1	1
3	6	2	0	5	11	2	1	9	11	1	2
3	7	2	1	5	12	2	2	9	12	1	3
3	8	2	2	6	6	1	0	10	10	1	0
3	9	3	0	6	7	1	1	10	11	1	1
3	10	3	1	6	8	1	2	10	12	1	2
3	11	3	2	6	9	1	3	11	11	1	0
3	12	4	0	6	10	1	4	11	12	1	1
4	4	1	0	6	11	1	5	12	12	1	0

2 in 2	1		4 in 4	1		6 in 6	1		8 in 8	1		10 in 10	1
2	4	2	4	8	2	6	12	2	8	16	2	10	20	2
2	6	3	4	12	3	6	18	3	8	24	3	10	30	3
2	8	4	4	16	4	6	24	4	8	32	4	10	40	4
2	10	5	4	20	5	6	30	5	8	40	5	10	50	5
2	12	6	4	24	6	6	36	6	8	48	6	10	60	6
2	14	7	4	28	7	6	42	7	8	56	7	10	70	7
2	16	8	4	32	8	6	48	8	8	64	8	10	80	8
2	18	9	4	36	9	6	54	9	9	72	9	10	90	9
3	3	1	5	5	1	7	7	1	9	9	1	11	11	1
3	6	2	5	10	2	7	14	2	9	18	2	11	22	2
3	9	3	5	15	3	7	21	3	9	27	3	11	33	3
3	12	4	5	20	4	7	28	4	9	36	4	11	44	4
3	15	5	5	25	5	7	35	5	9	45	5	11	55	5
3	18	6	5	30	6	7	42	6	9	54	6	11	66	6
3	21	7	5	35	7	7	49	7	9	63	7	11	77	7
3	24	8	5	40	8	7	56	8	9	72	8	11	88	8
3	27	9	5	45	9	7	63	9	9	81	9	11	99	9

RÉVIEW.

1. How do you write down the terms in this rule? 2. What next? 3. When there is no remainder? Proof. 4. What is the sign, and what does it mean? — Do you understand the table?

EXAMPLES.

Divisor 3)976432 dividend.

Quotient 325477—1 over, or }
　　　　　　×3+1　[rem. }

Proof　　976432

Explanation.—Begin and say, 3 is in 9, 3 times, which set down; then 3 is in 7 2 times and 1 over; this 1 is now supposed to be placed at the left of the next figure 6, which would be 16; then say 3 is in 16, 5 times and 1 over; then 3 is in 14, 4 times and 2 over; suppose 2 placed on the left of the next figure 3, which will make 23; then 3 is in 23, 7 times and 2 over; this 2 placed at the left of the next figure 2 is 22; then 3 is in 22, 7 times and 1 over—a remainder which place on the right with a *dash* (—1) or *cross* (+1). Then, to prove the work, multiply the quotient by the divisor and add in the remainder (1), and the product agrees with the dividend.

	(1.)	(2.)	(3.)	(4.)
Divisor	2)45245	2)673456	3)647045	4)862542
	22622+1	336728	215681+2	215635+2
	2×	2×	3×	4×
Proof	45245	673456	647045	862542

(5.)	(6.)	(7.)	(8.)	(9.)
2)648024	3)570624	4)592435	5)65497	6)345670

(10.)	(11.)	(12.)	(13.)	(14.)
7)414678	8)650784	9)479542	10)184567	11)164560

(15.)	(16.)	(17.)	(18.)	(19.)
12)186794	8)976042	9)740641	8)764512	7)640531

(20.) 4640213 ÷ 4 = quotient 1160053 + 1
(21.) 7564051 ÷ 5 = 1512810 + 1
(22.) 3245670 ÷ 6 = 540945
(23.) 4534214 ÷ 7 = 647744 + 6
(24.) 5640312 ÷ 8 = 705039
(25.) 9768532 ÷ 9 = 1085392 + 4
(26.) 8476523 ÷ 10 = 847652 + 3
(27.) 7864591 ÷ 11 = 714962 + 9
(28.) 8047654 ÷ 12 = 670637 + 10
(29.) 1976583 ÷ 7 = 282369
(30.) 1248234 ÷ 6 = 208039
(31.) 7640561 ÷ 5 = 1528112 + 1
(32.) 6404740 ÷ 4 = 1601185
(33.) 6470757 ÷ 9 = 718973

(34.) 4687096 ÷ 8
(35.) 7058407 ÷ 9
(36.) 8976452 ÷ 7
(37.) 6459876 ÷ 6
(38.) 7486541 ÷ 5
(39.) 7098462 ÷ 4
(40.) 3121462 ÷ 5
(41.) 1798471 ÷ 3

(42.) 6804970 ÷ 11
(43.) 8567984 ÷ 12
(44.) 7986459 ÷ 7
(45.) 8467965 ÷ 8
(46.) 1453642 ÷ 9
(47.) 7465981 ÷ 10
(48.) 6583549 ÷ 12

* (49.) A quantity of railroad stock, amounting to 24867 dollars, is to be equally divided among 9 stockholders; how much must each man receive for his share? *Ans.* 2763 dollars.

(50.) The expenses of a boarding-house amount to 7680 dollars in a year (12 months); how much is it for 1 month? *Ans.* 640 dollars.

(51.) If 7 persons sell property to the amount of 2247 dollars, how much will each man receive for his share? *Ans.* 321 dollars.

(52.) How many times are 9 cents contained in 633708 cents? *Ans.* 70412 times.

(53.) How many times is 11 contained in 780469511?

(54.) Twelve ships are to be laden with 91968 bushels of wheat; required the number of bushels for each ship. *Ans.* 7664 bushels.

(55.) Seven children agreed to divide equally the estate left by their father; the sum divided was 42287 dollars; required the share of each. *Ans.* 6041 dollars.

LONG DIVISION.

Long Division is generally used when the divisor is more than 12.

1. Write down the dividend, and draw curved lines at the right and left sides of the dividend, thus)80(; and place the divisor on the left-hand, as in Short Division.
2. See how often the divisor is contained in the least number of figures into which it can be divided, and set that number on the right of the dividend, and this is the multiplier.
3. Multiply the divisor by this figure in the quotient, and place the result under the figures in the left of the dividend, into which you are dividing.
4. Then subtract the result from the number directly over it, and set down the remainder, which must always be less than the divisor.
5. Bring down the next figure of the dividend, and place it on the right of the remainder; if this number is less than the divisor, place a cipher in the quotient, and bring down another figure, which increases the dividend, into which divide as before, and so continue until all the figures are brought down and divided.
6. If there be ciphers at the right of the dividend and divisor, you can omit an equal number of each, by placing a period, or comma, on the left of each, thus :—

$$1,00)10,00(10 \text{ Ans.}$$

When is Long Division used? 1. How do you write the terms? 2. After the terms are written, what is first to be done? 3. How do you multiply? 4. After you have multiplied the divisor by the quotient figure, what is the next operation? 5. What next? If this number is *less* than the divisor, what is to be done? Why do you bring down another figure? *Ans.* To increase or enlarge the dividend, so it may be equal to, or more than, the divisor. 6. When there are ciphers on the right of the divisor and dividend, what can be done? Having the dividend and quotient given, how is the divisor found? *Ans.* Divide the dividend by the quotient. If you have the divisor and quotient, how can you find the dividend? *Ans.* Multiply them together.

Divisor. Dividend. Quotient.

```
24 ) 6576 ( 274
     48        24 ×
    ---       ----
    177      1096
    168       548
    ---      ----
     96      6576  proof.
     96
    ---
Remainder 00
```

Explanation.—Consider how many times 24 is contained in 65; we know it to be twice, but not 3 times, for 3 times 24=72, which is more than 65; therefore, place 2 in the quotient and multiply 24 by 2, and the product is 48; write this under 65 and subtract it from 65, and 17 over. If there were no more figures in the dividend, the work would be finished; but there are two more figures to be divided. Place the next figure 7 on the right of 17—this makes a new dividend=177; now consider how many times 24 is contained in 177; suppose 7 times, write 7 in the quotient on the right of 2, then multiply 24 by 7, and the product is 168, which subtract from 177, and 9 is the remainder; bring down the next and last figure 6 on the right of 9=96; then 24 in 96, 4 times 24=96, and 0 remainder. Now it is evident, that if 24 is contained in 6576 just 274 times, that by multiplying the *quotient* and *divisor* together, that product must be equal to the *dividend*, if the operation has been correctly performed, and this we find to be the case. If the dividend is divided by the quotient, it will produce the divisor, because, if 24 is contained in 6576 just 274 times, 274 is contained in 6576 just 24 times—this is proved by multiplying those two numbers together, as shown above.

```
274)6576(24
    548
   ----
   1096
   1096
```

In this example, we have taken the first *quotient* for a *divisor*, and produced the first *divisor* 24.

Note.—There are several methods used by different authors to prove questions in the preceding rules; those given in this work are preferred by the most competent teachers. One method is by "*casting out the nines*," which, for many good reasons, is omitted in this work, as it would require more time and study, on the part of the pupil, than the extraction of the *cube root;* the process is uncertain, and can not be relied upon; it is a useless waste of time. A suitable example well explained, so as to be understood by the pupil, will convey more *instruction*, both in *theory* and *practice*, than a volume of *demonstrations*,

(4)

dissertations, axioms, &c., too freely introduced into our primary books.

(1.)	(2.)	(3.)	(4.)
16)478905(	34)567896(	50)658976(	71)65897(

(5.)	640997 ÷ 48 =	quotient 13354	+ 5
(6.)	840756 ÷ 78 =	10778	+ 72
(7.)	791605 ÷ 81 =	9772	+ 73
(8.)	415678 ÷ 97 =	4285	+ 33
(9.)	345678 ÷ 52 =	6647	+ 34
(10.)	141509 ÷ 105 =	1347	+ 74
(11.)	389152 ÷ 204 =	1907	+ 124
(12.)	567894 ÷ 115 =	4938	+ 24
(13.)	124689 ÷ 212 =	588	+ 33
(14.)	798642 ÷ 611 =	1307	+ 65
(15.)	418967 ÷ 720 =	581	+ 647
(16.)	489786 ÷ 514 =	952	+ 458
(17.)	600798 ÷ 321 =	1871	+ 207
(18.)	748960 ÷ 842 =	889	+ 422
(19.)	679845 ÷ 981 =	693	+ 12
(20.)	704536 ÷ 1001 =	703	+ 833

	Divisor.	Dividend.	Quotient.	Remainder.
(21.)	1262	70984657	56247	943
(22.)	1784	98705230	55328	78
(23.)	1496	63214563	42255	1083
(24.)	2582	58009768	22466	2556
(25.)	2904	79143584	27253	872
(26.)	3014	85367914	28323	2392

(When the divisor is the exact product of two or more figures.)

RULE.

1. Divide by each of those numbers separately instead of the whole divisor at once.

The actual remainder in questions worked by this rule is found as follows :—

2. Multiply the quotient by the divisor and subtract the product from the dividend, and the result will be the true remainder ; for if the product of the quotient and divisor, added to the remainder, be equal to the dividend, their product taken from the dividend must leave the remainder.

(27.) 6)68752 ÷ 144

6)11458 + 4

4)1909 + 4

477 + 1

```
477    quotient
144 ×  divisor

1908
1908
477

68688 product
```

```
       6
      6 ×
      —
      36
      4 ×
      —
     144
```

```
144)68752(477 quot't.
    576
    ———
    1115
    1008
    ————
    1072
    1008
    ————
proof  64 remainder.
```

```
68752  dividend
68688 — product

64   remainder
```

```
12)68752      12
   ————       12 ×
12)5729 + 4   ——
   ————       144
    477 + 5
    144 ×
    ————
    1908
    1908    68752
    477   — 68688
    ————   —————
prod. 68688 rem. 64
```

Note. — Any numbers may be used, which being multiplied together will produce the divisor; however, generally, Long Division is the most convenient.

(28.)	6876546 ÷ 24.	(4 × 6 = 24)
(29.)	7906405 ÷ 42.	(6 × 7 = 42)
(30.)	9854347 ÷ 56.	(7 × 8 = 56)
(31.)	4089765 ÷ 63.	(7 × 9 = 63)
(32.)	5487963 ÷ 99.	(11 × 9 = 99)
(33.)	6458764 ÷ 108.	(9 × 12 = 108)
(34.)	6509712 ÷ 144.	(12 × 12 = 144)
(35.)	7680457 ÷ 126.	(7 × 6 × 3)
(36.)	8490765 ÷ 252.	(4 × 9 × 7)
(37.)	6498764 ÷ 336.	(8 × 7 × 6)

Note. — It will be well for the pupil to solve the above examples by both methods, as one operation will prove the other (see example).

(38.) How many times is 224 contained in 89765880 ?

(39.) Divide 55272 dollars equally among 7 men.

Ans. 7896 dollars.

(40.) The crew of a ship consisting of 80 men are entitled to 17120 dollars prize-money; what is the share of each man?   *Ans.* 214 dollars.

(41.) What is the quotient of 79084128 divided by 64?

(42.) What number must be multiplied by 121 to produce 242?

(43.) General Taylor received 40500 pounds of flour for his army, which at that time consisted of 4500 men; how much flour did each man receive?   *Ans.* 9 pounds.

(44.) A worthy farmer bequeathed his estate of 18494 dollars to his 4 sons and 3 daughters, to be equally divided among them; required the share of each.

                 Ans. 2642 dollars.

(45.) Divide 1600 dollars equally among 10 men, 40 men, 160 men.

(46.) If 60 minutes make an hour, how many hours are are there in 45840 minutes?   *Ans.* 764 hours.

(47.) If 24 hours make a day, how many days are there in 8448 hours?   *Ans.* 352 days.

(48.) If 7 days make a week, how many weeks are there in 980 days?   *Ans.* 140 weeks.

(49.) If 12 months make a year, how many years in 604 months?

(50.) At 14 cents a pound, how many pounds can you have for 2100 cents?

(51.) If 365 days make a year, how many years in 2555 days?

(52.) If 16 ounces make 1 pound, how many pounds in 2272 ounces?   *Ans.* 142 pounds.

(53.) If 160 rods make an acre, how many acres in 12160 rods?   *Ans.* 76 acres.

(54.) If 12 inches make a foot, how many feet in 6744 inches?   *Ans.* 562 feet.

(55.) If 10 mills make 1 cent, how many cents in 87400 mills?   *Ans.* 8740 cents.

(56.) If 128 feet make a cord of wood, how many cords in 7808 feet?   *Ans.* 61 cords.

MISCELLANEOUS QUESTIONS,

APPLICABLE TO THE PRECEDING RULES.

(1.) If the multiplicand be 6408, and the multiplier 640, required the product.

(2.) If the divisor be 41, and dividend 697141, required the quotient.

(3.) Add 640, 784, 982, and divide the amount by 64 ?

(4.) A farmer has 3 tracts of land, one of 840 acres, one of 621 acres, and one of 1147 acres; he wishes to divide them into 8 farms for each of his children : how many acres will each receive ? *Ans.* 326 acres.

(5.) A rich merchant bequeathed his estate to his 4 children, which amounted to 256448 dollars; his 2 sons were to receive 128224 dollars, and the remainder of the estate was to be equally divided between his 2 daughters : how much did each receive, and did they all share alike ?

(6.) If you purchase 250 pounds of flour for 12 dollars, and sell it for 6 cents a pound, will you gain, or lose, and how much ? *Ans.* gain 3 dollars.

(7.) If you should purchase 475 acres of land for 9000 dollars, and sell it for 20 dollars an acre, would you gain, or lose, and how much ?

(8.) A merchant deposited in bank at one time 648 dollars; at another time, 742 dollars; a third time, 1400 dollars; in the meantime he has drawn 1152 dollars: how much has he remaining in the bank ? *Ans.* 1638 dollars.

(9.) A ship in sailing to a distant part of the world : from one port to another was 5849 miles, to another port 6420 miles, to another 8542 miles, which brought her to the island of Cuba, and thence home 924 miles; required the number of miles she sailed. *Ans.* 21735 miles.

(10.) Multiply 694 by 342, and divide the product by 720.

(11.) A farmer was indebted to a merchant 1192 dollars; he paid him 481 dollars, and then purchased goods to the amount of 614 dollars; how much did he then owe him ? *Ans.* 1325 dollars.

(12.) Purchased a horse for 75 dollars, 3 cows for 25 dollars each, 18 sheep for 2 dollars each, and sold them all for 180 dollars; did I gain, or lose, and how much ? *Ans.* lost 6 dollars.

(13.) Light passes from the sun to the earth, a distance of 95 millions of miles, in about 8 minutes; what distance does light move in a minute ? *Ans.* 11,875,000 miles per minute.

REMARKS ON CURRENCY.

THERE is but one currency known to our laws, and coined in the United States; that is, the eagle, half-eagle, quarter-eagle, dollar, half-dollar, quarter-dollar, dime, half-dime, cent, and half-cent. There are various kinds of foreign gold and silver in circulation throughout the Union, the value of which is established by acts of Congress; but book accounts, and all written contracts for money, must be in *dollars and cents*, or the currency of the Union. There are various methods of reckoning money in the several states of the Union, the cause of which may be traced to the condition of the country before the existence of our present excellent form of government. At that time the present United States were colonies of Great Britain, and subject to their laws; and their method of reckoning money in pounds, shillings, and pence, prevailed in this country, but was not uniform in all the states. In some states the dollar was estimated at six shillings—this would make one shilling sixteen and two thirds cents; in other states the dollar was estimated at eight shillings—this would make the value of one shilling twelve and a half cents; in some states, four shillings and sixpence, and in others, seven shillings and sixpence: but this neither reduced nor enhanced the real value of the dollar, which has always been the same. This method of reckoning is still persisted in by many, to the great inconvenience and trouble of the community—a system that should be discontinued and prohibited. The foreign silver coin mostly in circulation in this country is the Spanish real and half-real; or, the twelve-and-a-half-cent piece, and the six-and-a-quarter-cent piece: known by various names, as *shilling, levy, fip, bit*, &c., &c. This kind of money has been so long in circulation as to become worn, and very much reduced in value, and efforts are being made to recoin it into our own currency; this will soon rid the country of this very *trouble-some* and *unnatural* currency, much to the relief of the community, and then our money will be *constitutional*.

The money of the United States increases and decreases by *tens*, which is called DECIMAL; this makes it extremely easy to calculate: that is, tens of every lower denomination, or less value, make one of the next higher, and consequently one of every higher makes ten of the next lower: which may be seen by the following:—

TABLE I.

10 mills make, or equal, 1 cent   10 mills, m. mill.
10 cents   "   "   1 dime   100   "   d. dime.
10 dimes   "   "   1 dollar 1000   " $, or D. dollar.
10 dollars "   "   1 eagle 10000   "   E. eagle.

AMERICAN COINS.

The Mill is not a coin, but the tenth part of a cent.

The Half-Cent is a copper coin, 200 being equal to one dollar.

The Cent is a copper coin, 100 being equal to one dollar.

The Half-Dime is the smallest silver coin, being equal in value to 5 cents, or 20 to a dollar.

The Dime is a silver coin, equal to 10 cents, or 10 to a dollar.

The Quarter-Dollar is a silver coin, =25 cents, or 4 to a dollar.

The Half-Dollar is a silver coin, =50 cents, or 2 to a dollar.

The Dollar is the largest silver coin, =100 cents, or 10 =1 eagle.'

The Quarter-Eagle is the smallest gold coin, =2 dollars 50 cents.

The Half-Eagle is a gold coin, =5 dollars; 2 =1 eagle.

The Eagle is the largest gold coin, =10 dollars.

TABLE II.

Thousands of dollars.	Hundreds of dollars.	Eagles, or tens of dollars.	Dollars.	Dimes, or tens of cents.	Cents, or tens of mills.	Dollars.	Cents.	Parts.			Parts, or decimals.
		1	.	2	3	1 and	.23		or	123	
	1	3	.	1	2	13	.12			1312	
	4	2	.	0	9	42	.09			4209	
6	7	2	.	1	8	672	.18			67218	
5	6	9	.	0	4¾	569	.04 and ¾			56904	.75
7 0	1	5	.	4	2½	7015	.42	½		701542	.50
2 1	4	6	.	8	1¼	2146	.81	¼		214681	.25

OF FRACTIONS.

Before proceeding further, it will be necessary to give a short explanation of fractions, as no calculations of importance can be made without their application. But as this volume is intended as an introduction, and as this part of arithmetic is explained at large in the other volume, the examples in this work will be limited to mere simple fractions, sufficient for *practical* calculations.

Fractions are of two kinds, either *vulgar* or *decimal*, and mean a part or parts of *one* or unity; a whole is called an integer; but a part, or some parts, of an integer are denoted by figures, and in vulgar fractions are written with a line drawn between the numerator and the denominator, thus: $\frac{1}{4}$ numerator, denominator, $=$ one fourth; $\frac{1}{2}$ one half; $\frac{3}{4}$ three fourths; $\frac{7}{8}$ seven eighths, &c.

Decimal signifies ten, because decimal fractions *increase* and *decrease* in a tenfold ratio, and is a part of a unit or whole number, and is distinguished by a comma (,) or period (.) placed on the *left* of the figure or figures, thus .5, or thus $\frac{5}{10}$, *five tenths*, or one half; .25 $= \frac{25}{100}$, *twenty-five hundredths*, or one quarter. A fraction is of the same *value*, whether expressed in the form of a vulgar fraction or decimal. We will take 1 dollar as the *unit*, in our currency, and give the several parts in vulgar and decimal fractions :—

VULGAR FRACTIONS.

1 D. $\frac{1}{16}=6\frac{1}{4}$ cts.; $\frac{2}{16}=\frac{1}{8}=12\frac{1}{2}$ cts.; $\frac{3}{16}=18\frac{3}{4}$ cts.; $\frac{4}{16}=\frac{1}{4}=$ 25 cts.; $\frac{5}{16}=31\frac{1}{4}$ cts.; $\frac{6}{16}=\frac{3}{8}=37\frac{1}{2}$ cts.; $\frac{7}{16}=43\frac{3}{4}$ cts.; $\frac{8}{16}=\frac{4}{8}=\frac{1}{2}=50$ cts.; $\frac{9}{16}=56\frac{1}{4}$ cts.; $\frac{10}{16}=\frac{5}{8}=62\frac{1}{2}$ cts.; $\frac{11}{16}=68\frac{3}{4}$ cts.; $\frac{12}{16}=\frac{3}{4}=75$ cts.; $\frac{13}{16}=81\frac{1}{4}$ cts.; $\frac{14}{16}=\frac{7}{8}=87\frac{1}{2}$ cts.; $\frac{15}{16}=93\frac{3}{4}$ cts.; $\frac{16}{16}=100$ cts.$=1$ D.

DECIMAL EXPRESSIONS.

$6\frac{1}{4}$ cents $= \underline{62.5}$ mills $= \underline{6\frac{1}{4}} = 6.25$ } cents.
1 dollar $= 1000$ mills $\quad 100 \quad .6\frac{1}{4}$

$12\frac{1}{2}$ cents $= \underline{125}$ mills $= \underline{12\frac{1}{2}} = 12.5$ } cents.
1 dollar $= 1000$ mills $\quad 100 \quad .12\frac{1}{2}$

$18\frac{3}{4}$ cents $= \underline{187.5}$ mills $= \underline{18\frac{3}{4}} = 18.75$ } cents.
1 dollar $= 1000$ mills $\quad 100 \quad .18\frac{3}{4}$

25 cents $= \underline{250}$ mills $= \underline{25} = 25$ } cents.
1 dollar $= 1000$ mills $\quad 100 \quad .25$

$31\frac{1}{4}$ cents $= 312.5$ mills $= 31\frac{1}{4} = 31.25$
1 dollar $= \overline{1000}$ mills $\overline{100}$ $31\frac{1}{4}$ } cents.

$37\frac{1}{2}$ cents $= 375$ mills $= 37\frac{1}{2} = 37.5$
1 dollar $= \overline{1000}$ mills $\overline{100}$ $37\frac{1}{2}$ } cents.

$43\frac{3}{4}$ cents $= 437.5$ mills $= 43\frac{3}{4} = 43.75$
1 dollar $= 1000$ mills 100 $.43\frac{3}{4}$ } cents.

50 cents $= 500$ mills $= 50 = 5 = .50$
1 dollar $= \overline{1000}$ mills $\overline{100}$ $\overline{10}$ $\frac{1}{2}$D. $= .5$ } cents.

$56\frac{1}{4}$ cents $= 562.5$ mills $= 56\frac{1}{4} = 56.25$
1 dollar $= \overline{1000}$ mills $\overline{100}$ $.56\frac{1}{4}$ } cents.

$62\frac{1}{2}$ cents $= 625$ mills $= 62\frac{1}{2} = 62.5$
1 dollar $= \overline{1000}$ mills $\overline{100}$ $.62\frac{1}{2}$ } cents.

$68\frac{3}{4}$ cents $= 687.5$ mills $= 68\frac{3}{4} = 68.75$
1 dollar $= \overline{1000}$ mills $\overline{100}$ $.68\frac{3}{4}$ } cents.

75 cents $= 750$ mills $= 75 = .75$
1 dollar $= \overline{1000}$ mills $\overline{100}$ $\frac{3}{4}$ D. } cents.

$81\frac{1}{4}$ cents $= 812.5$ mills $= 81\frac{1}{4} = 81.25$
1 dollar $= \overline{1000}$ mills $\overline{100}$ $.81\frac{1}{4}$ } cents.

$87\frac{1}{2}$ cents $= 875$ mills $= 87\frac{1}{2} = 87.5$
1 dollar $= \overline{1000}$ mills $\overline{100}$ $87\frac{1}{2}$ } cents.

$93\frac{3}{4}$ cents $= 937.5$ mills $= 93\frac{3}{4} = 93.75$
1 dollar $= 1000$ mills 100 $.93\frac{3}{4}$ } cents.

100 cents $= \dfrac{10}{10} = \dfrac{1}{1} = 1$ D.; $.25 = \frac{1}{4}$, $.50 = \frac{1}{2}$, $.75 = \frac{3}{4}$.
1 dollar

The preceding division is in accordance with the Spanish "*real and half-real:*" that is, the shilling and sixpence. The following is *our coin :*—

1 mill $= \frac{1}{10}$ of a cent; 1 cent $= \frac{1}{10}$ of a dime; 1 dime $= \frac{1}{10}$ of a dollar; 1 dollar $= \frac{1}{10}$ of an eagle. Or, 1 cent $= \frac{1}{100}$ of a dollar; 10 cents $= \frac{10}{100}$; 20 cents $= \frac{20}{100}$; 30 cents $= \frac{30}{100}$; 40 cents $= \frac{40}{100}$; 50 cents $= \frac{50}{100}$; 60 cents $= \frac{60}{100}$; 70 cents $= \frac{70}{100}$; 80 cents $= \frac{80}{100}$: 90 cents $= \frac{90}{100}$.

The superiority of this currency is "self-evident."

AMERICAN CURRENCY, OR U. STATES MONEY.

ADDITION.

RULE.—1. Write dollars under dollars, cents under cents, mills under mills, and point under point, and add them up as whole numbers; write a period or comma in the amount, directly under those above. 2. When dimes are in the given sum, but one place is required; but for cents, two places. If the cents be less than ten, a cipher must be prefixed on the left-hand of the figure which expresses them, thus .08, eight cents. 3. There may be a small space left between the denominations and the points omitted, thus :—

D. c. m.	D. c. m.	D. c. m.	D. c.	D. d. c. m.
5 25 5	or 5.25.5	or 5,25,5	or 5.25¼	or 5,2,5,5.

Proof, the same as Addition of whole numbers.

EXAMPLES.

	(1.)			(2.)			(3.)			(4.)	
D.	c.	m.	D.	c.	m.	D.	c.	m.	D.	d.	c.
4	21	4	74.	82.	7	141	, 20	, 5	72	8	3
6	07	6	38.	41.	6	132	, 72	, 0	44	6	5
3	18	5	64.	05.	7	684	, 31	, 5	56	7	5½
5	06	2	12.	64.	1	78	, 01	, 7	62	4	8¼
17	74	9	07.	50.	5	01	, 41	, 3	34	5	5¼

=1 cent to 5=6

	(5.)				(6.)				(7.)		(8.)		
E.	D.	d.	c.	m.	E.	D.	c.	m.	D.	c.	E.	D.	c.
11	4	5	3	6	147	8	9	6	145	, 62	21	7	22
7	6	4	2	1	284	6	7	9	381	, 74	14	6	42
5	7	2	1	0	821	6	7	8	621	, 31	13	4	08
3	5	6	7	8	200	5	6	7	205	, 61½	7	6	40
9	7	6	4	2	481	2	6	5	140	, 31¾	8	6	92

(205, 61½ and 140, 31¾ braced } 1)

(9.) Add 11D. 62c. 7m.; 18D. 79c. 8m.; 148D. 08c. 7m.; 195D. 79c. 7m.; 10D. 74c.; together. *Ans.* 385D. 04c. 9m.

(10.) 149D. 64c. 8m.+948D. 65c. 7m.+479D. 84c. 9m.+ 1001D. 80c.+1460D. 70c.   *Ans.* 4040D. 65c. 4m.

(11.) A merchant has the following charges in his ledger, namely : to A. 150D. 62½c. ; to B. 641D. 84c. ; to C. 722D. 92c. 5m. ; to D. 218D. 21c. : required the amount.

Ans. 1733D. 60c.

(12.) A drover purchased a drove as follows: for cows 784D. 62c., for calves 221D. 81¼c., for sheep 321D. 21¼c., 1 colt for 75D., and 3 horses for 82D. each; required the cost of the drove. *Ans.* 1648D. 64¾c.

(13.) A farmer sold wheat for 584D. 78c.; rye, 621D. 92c. 5m.; oats, 181D. 92c.; corn, 491D. 50c.; potatoes, 50D.; hay, 74D. 71¼c.; veal, 5D.; and flax, 2D. 50c.: required the amount. *Ans.* 2012D. 33¼c.

SUBTRACTION.

RULE.—1. Write down the numbers the same as in Simple Subtraction, observing that the points stand directly under each other, and dollars under dollars, cents under cents, &c. 2. Then subtract the same as in whole numbers, and place the points in the remainder under those above.

Proof, as Subtraction of whole numbers.

	(1.) D.	c.	m.		(2.) D.	c.	m.		(3.) D.	c.	m.
	65	71	5		689	78	4		64	80	0
	21	84	4		546	29	5		29	90	8
Rem.	43	87	1		143	48	9		34	89	2
Proof	65	71	5		689	78	4		64	80	0

	(4.) D.	c.		(5.) D.		(6.) D	d.	c.	m.		(7.) E.	D.	c.
	789	60		7984		784	1	4	2		47	8	62
	540	74		640		520	8	2	4		25	6	47
Rem.	248	86		7344									
Proof	789	60		7984									

(8.) E.	D.	d.	c.	m.	(9.) D.	c.	m.	(10.) D.	c.	m.
98	7	6	4	2	140	29	0	940	00	0
41	9	8	5	1	78	00	9	642	84	5

(11.) From 194D. 78c. take 50D. 90c. 8m.

(12.) From 75cts. take 8 mills.

(13.) From 19 eagles take 19D. 19c. 9m. (19E.=190D.)
(14.) From 16D. take 1E. 1D. 1c. 1m.
(15.) From 1798D. take 105D. 90c. 7m.
(16.) A merchant deposited in bank 500 eagles; he drew out at one time 1892D. 68c. 5m., at another 1141D. 62c.: how much has he in bank?   *Ans.* D1965.69.5.
(17.) A farmer sold a miller his wheat for 687D., and received cash in part payment, 487D. 92c. 4m., and a note for the balance due for the wheat; required the amount of the note.   *Ans.* D199.07.6.
(18.) The collector on the canal at Easton received in tolls, in one week, 2847D. 95c. 5m.; in another week, 3421D. 47c.; of the above amount he paid an order of 3694D. 87c. 5m.; how much has he remaining of the amount collected?   *Ans.* D2574.55.

REVIEW.

1. How do you write sums in Addition of United States money? 2. How many places of figures are required for dimes? How many for cents? Proof. 1. How do you write down numbers in Subtraction? How do you subtract? Proof.

MULTIPLICATION.

RULE.—Multiply as in common Multiplication, and point off one figure to the right-hand of the product for mills, the next two for cents, the remainder will be dollars. If there are no mills in the quantity multiplied, point off two places to the right; the remainder will be dollars.

Proof, as Simple Multiplication, or more correctly by Division.

EXAMPLES.

(1.) What will 12 yards of tape cost at 25c. per yard?
25
12 ×
D3.00 *Ans.*

(2.) What will 9 yards of calico cost at 37c. 5m. (37½c.) per yard?
37.5
9 ×
D3.37.5

(3.) What will 7 yards of riband cost at 62c. 5m. (62½c.)
per yard?
$$62.5$$
$$7\times$$
$$\textit{Ans. } D\overline{4.37.5}$$

(4.) What cost 8 yards of linen at 75c. per yard?
$$75\times8=D6.00 \ \textit{Ans.}$$

(5.) What cost 12 yards of cloth at 2D. 25c. per yard?
$$2.25\times12=D27.00 \ \textit{Ans.}$$

(6.) What will 20 yards cost at D1.12.5 per yard?
Ans. D22.50c.

(7.) What will 18¾ pounds of butter cost at 16 cents per
pound? *Ans.* D3.00.
Thus: $18.75\times16=D3,00,00$; count off two places for .75,
and two for cents.

(8.) What will 140 bushels of wheat cost at D1.92 per
bushel?

(9.) What will 1198 pounds of sugar cost at 9 cents per
pound? *Ans.* D10.78.2.

(10.) What will 7 tons of coal cost D4.25 per ton?
Ans. D29.75.

(11.) What will 55 pounds of tea cost at D1.12.5 per
pound? *Ans.* D61.87.5.

(12.) What will 40 barrels of flour cost at D9.25 per barrel? *Ans.* D370.00.

(13.) How much can you earn in 95 days at 62c. 5m. per
day? *Ans.* D59.37.5.

(14.) What will 1250 pounds of cheese cost at 7c. 5m. per
pound? *Ans.* D93.75.

(15.) Potatoes are worth 87c. 5m. per bushel; what will
50 bushels cost?

(16.) How much will your board cost for 17 weeks at
D2.25 per week?

(17.) How much will 158 gallons of oil cost at D1.37.5
per gallon? *Ans.* D217.25.

(18.) What will 25 cords of wood cost at D3.75c. per cord?
Ans. D93.75.

(19.) How much will 70 sacks of salt cost at D5.62 per
sack? *Ans.* D39.34.

(20.) What will 74 boxes of raisins cost at D2.60 per
box? *Ans.* D192.40.

(21.) How much will 97 yards carpeting cost at D2.17
per yard? *Ans.* D210.49.

(22.) What will 191 bushels of oats cost at 52c. per
bushel? *Ans.* D99.32.

(5)

(23.) What will 74 acres of land cost at D62.50 per acre?
Ans. D4625.00.

(24.) Paid for a silk handkerchief D1.37.5; what would 50 cost?
Ans. D68.75.

(25.) What will it cost to fill a feather-bed with 25 pounds of feathers worth 67½c. per pound?
An. D16.87.5.

(26.) If the price of a musket was D13.50 for the volunteers in Mexico, what did it cost to furnish a regiment of 700 men?

(27.) What cost 70 pounds of lard at 12½c. per pound?
Ans. D8.75.

(28.) What cost 695 pounds of pork at 9c. 5m. per pound?
Ans. D66.02.5.

(29.) When oysters are worth 60c. per hundred, what will 3000 cost?

(30.) What cost 498 pounds of ham at 11½c. per pound?
Ans. D57.27c.

(31.) What will 280 pounds of iron cost at 13c. 7m. per pound?
Ans. D38.36c.

(32.) What will 95 pounds of flax cost at 9c. 3m. per pound?
Ans. D8.83.5.

(33.) What will 600 yards of Irish linen cost at D2.25c. per yard?

(34.) Find the amount of the following bills:—

New-York, Nov. 10, 1847.

Mr. Peter H. Wilmot

Bought of Henry Quinn:

12 pounds of loaf-sugar at	18 cts.	
16 pairs of morocco shoes .	1:12	
5 hats . .	3:75	
70 pounds of coffee .	:12 ½	
8 pairs of ladies silk hose	1:75	
5 yards of fine black cloth	9:25	

D107.83

Received payment,

Henry Quinn.

(35.)   Baltimore, Nov. 12, 1847:

Mr. Samuel Semenar

                    Bought of Wallace & Grant:

. 40 yards of flannel at .   :75 cts.
  1 set of dinner plates .   4.80
  87 gallons molasses   .   .60
  65 pounds of cheese   .   : 7 1/2
  25 pounds of raisins   .   :16
  42 yards of sheeting   .   :76
   4 pounds of indigo   .   2:75

  1. shawl   .   .   40:00
  20 yards cambric-muslin .   .60

                              _____
                              D190:79.5

  Received payment,

                    Wallace & Grant.

                    _____

DIVISION.

Rule.—1. Reduce the quantity to be divided to mills, by
annexing ciphers, and proceed to divide as in common Division, separating the three right-hand figures in the quotient.
2. The left-hand figures will then represent the dollars; the
first two on the right, the cents; and the last on the right,
the mills.

Proof, as Simple Division.

EXAMPLES.

(1.) Paid D63.12.5 for 25 yards of cloth; how much was it per yard?

(2.) Paid D100 for 80 bushels of wheat; what cost 1 bushel?

OPERATION.

Yards. D. c. m. D. c. m.
25) 63,12,5 (2,52,5 *Ans.*
   50
  ——
  131
  125
  ——
   62
   50
  ——
  125
  125
  ——
    0

Bushel. D.   D. c. m.
80) 100,00,0 (1,25,0 *Ans.*
  80
  ——
  200
  160
  ——
  400
  400
  ——
   0

(When the answer is required in money, the dividend must be money.)

(3.) Divide D400 equally among 15 men.
        Ans. D26,66,6$\frac{6}{10}$ mills, or $\frac{2}{3}$ of a mill.

(4.) Three men received D150.75; what is the share of each?

(5.) Paid D9.00 for 30 yards sheeting; how much was it per yard?

(6.) Received D25 for working 15 days.; how much was it per day?   *Ans.* D1,66,6$\frac{2}{3}$ mills.

(7.) A ship's crew of 70 men received in prize-money D15400; how much was it for each man?   *Ans.* D220.00.

(8.) A drover paid D105 for a flock of 60 sheep; how much did he pay per head?   *Ans.* D1.75.

(9.) How many half-dimes in D59.25?   *Ans.* 1185.

(10.) How many times are 12c. 5m. contained in D75,75?
                           Ans. 606.

(11.) Paid D125 for 37½ yards of broadcloth; what cost 1 yard?   D.
                               37,5)125,00,0(3,33,3+

(12.) How many dollars in 1101 quarters?   *Ans.* 276.

(13.) Wheat-flour is now selling at D8,50 per barrel of 196 pounds; how much is that per pound?
                        Ans. 4cts. 3m.+

(14.) How many dollars in a thousand dimes? in 500 eagles?

(15.) Oats are worth 70 cents a bushel; how many bushels can I purchase with 60 dollars? *Ans.* 85¾.

(16.) A farmer received 750 bushels of rye from 30 acres of land; how many bushels were raised on 1 acre? *Ans.* 25.

(17.) If a schoolboy should spend 20 cents a-day for nuts, ale, and cigars, how long would it take him to dispose of 100 dollars?

(18.) If you can manage to smoke daily 6 half-Spanish cigars at 2 cents each, how many days will it require to waste 24 dollars? and if you were employed at labor by the month, at 4 dollars a month, how many months' wages of your labor would be required to pay the bill?

Ans. to spend 24D. 200 days; and 6 months of labor.

(19.) A father divided an estate of 30000 dollars equally among 7 children; required the share of each.

Ans. D4285,71,4⅔ mills.

(20.) Paid D500 for 125 pounds of tea; what was the price of 1 pound?

(21.) Paid 18 dollars for 112 pounds of sugar; required the price of 1 pound.

(22.) Bought 400 bushels of oats for 300D.; how much did 1 bushel cost?

(23.) Bought 180 bushels of salt for 80D. 50c.; how much was it per bushel? 4cts. 7m.+

(24.) If your income is 562 dollars, 50cts. in a year, how much is that in a day?—(365 days.) *Ans.* D1,54,1.+

(25.) Bought 120 yards of superfine broadcloth for D850; how much must I sell it for per yard to gain D100 ?

Ans. D7,91,6.+

(26.) Paid D75,50 for a piece of silk containing 20 yards; required the cost of 1 yard. *Ans.* D3,77,5.

(27.) A grocer paid 345 dollars for 160 boxes of oranges, how much did he pay for 1 box? *Ans.* D2,15,6.+

(28.) A manufacturer purchased 20 bales of cotton, 300 pounds in each bale, for which he paid D750; how much was it per pound? *Ans.* 12c. 5m.

(29.) I have 3 pieces of cloth, each piece 18 yards, which cost me D256,50c.; I will sell it for D4,75 per yard: shall I receive as much as the first cost?

(30.) Mr. Brown died, leaving an estate valued at D25650; he directed that D840 should be given to the church, and D560 given to the poor, and that the remainder should be equally divided among his 7 children; required the share of each child, *Ans.* D3464,28,5.+

(5*)

APPLICATION OF THE PRECEDING RULES.

(1.) A farmer sold a barrel of pork for D19.50, and received in payment 6 pounds of tea at D1.25 per pound, and 1 hat at D3.50, and the balance in money; how much money did he receive?   *Ans.* D8.50.

(2.) A trader bought 250 sheep at 4 dollars per head, and paid for them in corn at 50 cts. per bushel; how many bushels did it require to pay for the sheep?   *Ans.* 2000.

(3.) Two steamboats on the Hudson start from New York for Albany, the distance being 160 miles; one moves at the rate of 20 miles an hour, and the other 18 miles an hour; the first boat will reach Albany in 8 hours: how far from Albany will the second boat be at the same time?
Ans. 16 miles.

(4.) A drover, who has 364 dollars, wishes to buy all the cows he can pay for at 16 dollars a head, and then pay out the remainder in sheep at 2 dollars a-head; how many of each must he purchase?   *Ans.* 22 cows, 6 sheep.

(5.) A trader purchased 25 tons of hay at 16 dollars a ton; at how much must he sell it per ton to gain 100 dollars?
Ans. 20 dollars.

(6.) What number must be multiplied by 40, in order that the product shall be 1000?   *Ans.* 25.

(7.) What number must be divided by 18, in order that the quotient shall be 12?

(8.) Three men purchased a ship; the first paid D5692,50, the second paid just twice as much as the first, and the third as much as both the first and second; required the cost of the ship?

(9.) A grocer purchased a cask of sugar weighing 1400 pounds, paid for transportation D4.50, and for the sugar D154, and he wishes to make D25,50; how much per pound must he sell it for?   *Ans.* 13c. $1\frac{6}{14}$ mills

(10.) Add together 3 dollars, 4 dollars and 6 cents, 7 dollars, 8 dollars and 2 cents, 5 dollars, 9 dollars and 8 cents, 5 dollars.

(11.) The population of the world has been estimated to be as follows: North America, twenty-six millions; South America, twelve millions; Europe, two hundred and twenty millions; Asia, five hundred millions; Africa, thirty-eight millions; Australia, four millions: what is the whole number?   *Ans.* 800 millions.

(12.) What will it cost to construct a railway from New York to Albany, a distance of 160 miles, at 12600 dollars per mile? *Ans.* D2016000.

(13.) The Secretary of War receives a salary of 6000 dollars per annum; how much is that daily? *Ans.* D16,43,8.+

(14.) If the salary of a clerk is 1650 dollars a-year, and he spends 3 dollars a-day, how much will he save? *Ans.* D555.

(15.) Purchased 7 chests of tea, each chest contained 75 pounds, at 1D. 62½c. per pound; required the cost. *Ans.* D853,12,5.

(16.) A. bought 40 yards of broadcloth at D4,25 per yard, and 35 yards at 5 dollars per yard; which piece cost the most, and how much did they both cost? *Ans.* 2d piece 5D. most; D345 cost of both.

(17.) Bought 5 hogsheads of molasses, each containing 63 gallons, for 50 dollars, and sold it for 20c. per gallon; how much did I make by the bargain? *Ans.* 13 dollars.

(18.) Bought 50 bushels of wheat for D1.80 per bushel, to be paid in rye at D1.50 per bushel; how many bushels of rye will pay for the wheat? *Ans.* 60.

(19.) If ⅙ of a sack of salt cost D4,62,5, how much did the whole sack cost? *Ans.* 4,62,5 × 6 = D27,75.

(20.) If $\frac{1}{10}$ of a dollar will pay for a yard of riband, how many yards can you have for D3,25?

(21.) If ⅕ of a hogshead of sugar is worth D9,24, what is the whole worth? *Ans.* D46,20.

(22.) A farmer has 99 bushels of corn; he will give his son ⅓ of it: how many bushels will each have? *Ans.* 33, 66.

(23.) When butter is ⅛ of a dollar a pound, how many pounds can you have for 15 dollars? *Ans.* 120.

REVIEW.

How will you multiply in dollars, cents, and mills? If there are no mills in the question? Proof. 1. Division: what is the first process in Division? 2. After you have divided, how will you separate the dollars, cents, and mills? How can you prove Division? Can you divide D25,25c. equally among 4 men?

TABLES OF WEIGHTS AND MEASURES.

AVOIRDUPOIS WEIGHT is the weight generally used in weighing groceries, and all coarse and heavy commodities.

Marked

16 drams (dr.) - make -	1 ounce -	-	oz.
16 ounces - - " -	1 pound -	*	lb.
28 pounds - - " -	1 quarter -	-	qr.
4 quarters (112 lbs.) " -	1 hundred-weight		cwt.
20 hundred-weight " -	1 ton -	-	T.
25 pounds (net) - " -	1 quarter -	-	qr. h.
4 quarters (100 lbs.) " -	1 hundred -	-	hund.

TROY WEIGHT is used in weighing gold and silver.

24 grains (gr.) - make -	1 pennyweight -	dwt.
20 pennyweights - " -	1 ounce - -	oz.
12 ounces - - " -	1 pound - -	lb.

APOTHECARIES' WEIGHT is used for compounding medicines, but not in selling them.

20 grains (gr.) - make -	1 scruple - -	Ә
3 scruples - - " -	1 dram - -	3
8 drams (drachms) " -	1 ounce - -	℥
12 ounces - - " -	1 pound - -	℔

CLOTH MEASURE is used in measuring cloth, calico, ribands, &c.

4 nails (na.) - make -	1 quarter - -	qr.
4 quarters - - " -	1 yard - -	yd.
5 quarters - - " -	1 English ell -	E. E.
5 quarters - - " -	1 French ell	Fr. E.
3 quarters - - " -	1 Flemish ell	Fl. E.

LIQUID MEASURE is used for beer, wine, vinegar, molasses, &c.

4 gills (gi.) - make -	1 pint -	pt.
2 pints - - " -	1 quart - -	qt.
4 quarts - - " -	1 gallon -	gall.
16 gallons - - " -	1 half-barrel	hf. bar.
31¼ (31.5) gallons " -	1 barrel -	bar.
42 gallons - - " -	1 tierce -	tier.
63 gallons - - " -	1 hogshead	hhd.
84 gallons - - " -	1 puncheon	punch.
2 hogsheads - - " -	1 pipe or butt	p. or b.
2 pipes or butts - " -	1 tun -	T.

DRY MEASURE is used in measuring grain, salt, &c.

Marked

2 pints (pt.)	-	-	make	1 quart - -	qt.
8 quarts -	-	-	"	1 peck - -	pk.
4 pecks -	-	-	"	1 bushel - -	bu.

LONG MEASURE is applied to length, distance, &c.

3 barley-corns (bar-c.) make	1 inch - -	in.
12 inches - - "	1 foot - -	ft.
3 feet - - "	1 yard - -	yd.
5½ (5.5) yards, or 16½ feet "	1 rod, pole, perch	r. p.
40 poles - - "	1 furlong - -	fur.
8 furlongs (1760 yds.) } "		
(320 poles) } "	1 mile -	M. mi.
69½ statute or } miles "		
60 geographic }	1 degree -	- deg.
360 degrees (the earth's } "		
circumference) - }	1 circle - - -	cir.

LAND or SQUARE MEASURE is used in measuring land, flooring, &c.; it has respect to length and breadth, but not to depth.

144 square inches (sq. in.) make	1 square foot sq. ft.
9 square feet - - "	1 square yard sq. yd.
30¼ square yards (or 272¼ } "	
square feet) }	1 squ. rod or pole r. p.
40 square rods - - "	1 square rood - R.
4 square rods - - "	1 square acre - A.
640 square acres (one sec- } "	
tion of land) }	1 square mile M. mi.

CUBIC MEASURE is used in measuring solid bodies, such as stone, timber, wood, &c. A cube is a solid of 6 equal sides.

1728 inches - - make	1 foot - - c. ft.
27 cubic feet - - "	1 yard - - c. yd.
40 feet of round timber "	1 ton - - T.
50 feet of hewn timber "	1 ton - - T.
16 cubic feet - - "	1 foot of wood - ft. w.
8 feet of wood - "	1 cord of wood C.
128 cubical feet - - "	1 cord of wood C.

MOTION or CIRCLE MEASURE is used by Navigators and Astronomers.

				Marked
60 seconds (″)	-	- make	1 minute - -	′
60 minutes	-	- "	1 degree - -	○
30 degrees	-	- "	1 sign - -	sig.
12 signs (360 deg.)	-	- "	1 revolution, or } circle,	rev.

TIME is naturally divided into days, by the revolution of the earth upon its axis; and into years, by the revolution of the earth around the sun.

60 seconds (sec.)	-	make	1 minute - -	m.
60 minutes	-	"	1 hour - -	h.
24 hours	- -	- "	1 day - -	da.
7 days	- -	- "	1 week - -	w.
4 weeks	- -	- "	1 month (lunar)	mo.
12 months } 52 weeks } 365¼ days }	- -	"	1 year - -	yr.

Every 4th or leap-year has 366 days.

The year is also divided into 12 calendar months, as follows :—

1st	month,	January,	has 31	days.
2d	month,	February,	28	"
3d	month,	March,	31	"
4th	month,	April,	30	'
5th	month,	May,	31	'
6th	month,	June,	30	'
7th	month,	July,	31	'
8th	month,	August,	31	"
9th	month,	September,	30	"
10th	month,	October,	31	"
11th	month,	November,	30	"
12th	month,	December,	31	"

For convenience, let the pupil commit the following to memory :—

" Thirty days hath September,
April, June, and November;
February twenty-eight alone;
All the rest have thirty-one."

N. B. In bissextile, or leap-year, February has 29 days.

PAPER.   *Marked*

24 sheets (sh.)	-	- make	1 quire	- -	qr.
20 quires	- -	- "	1 ream	- -	re.
2 reams	- -	- "	1 bundle	- -	bun.

PARTICULARS.

12 single things	-	- make	1 dozen	- -	doz.
12 dozen	-	- "	1 gross	- -	gr.
12 gross (144 doz.)	-	- "	1 great gross	G.	gro.
20 single things	-	- "	1 score	- -	sco.
5 scores	-	- "	1 hundred	-	hund.

REVIEW OF THE TABLES.

Avoirdupois Weight.—Recite the table. What is the use of this weight? How many drams in 3 pounds? How many ounces in 5 pounds? How many hundred-weight in 20 quarters? How many hundred-weight in 4 tons? How many pounds in 3 quarters?

Troy Weight.—What is weighed by Troy Weight? Recite the table. How many ounces in 4 pounds? How many pennyweights in 3 ounces? How many grains in 3 pennyweights? How many in 4?

Apothecaries' Weight.—How is Apothecaries' Weight used? Recite the table. How many drams in 25 scruples? How many scruples in 5 drams? How many ounces in 7 pounds? How many ounces in 8 pounds? How many ounces in 12 pounds?

Cloth Measure.—What is the use of this measure? How many nails in 7 quarters? How many quarters in 9 yards? How many yards in 16 quarters? How many quarters in 20 nails?

Liquid Measure.—When is this measure applied? Recite the table. In 6 gallons of molasses, how many quarts? How many pints in 19 quarts of vinegar? How many gills in 2 quarts? How many gills in 11 pints of milk? How many pints in 28 quarts? How many gills in 7 pints? In 12 pints?

Dry Measure.—What is the use of Dry Measure? Recite the table. How many quarts in 5 pecks? How many pecks in 11 bushels? How many bushels in 20 pecks? How many pecks in 32 quarts? How many quarts in 3 bushels?

Long Measure.—When is this measure used? Recite the table. How many barley-corns in 11 inches? How many inches in 9 feet? How many inches in 2 yards? How many rods in 3 furlongs? How many miles in 8 furlongs? How many inches in 4 feet? In 8 inches how many barley-corns? In 12 inches?

Land or Square Measure.—When is this measure applied? Recite the table. How many square feet in 5 square yards? How many square inches in 2 square feet? How many rods in 3 roods? How many acres in 24 roods? In 28 acres?

Cubic Measure.—What is the use of this measure? Recite the table. How many feet in half a cord of wood? In a quarter of a cord? In three quarters of a cord? In $\frac{1}{8}$ of a cord?

Motion or Circle Measure.—By whom is this measure used? Recite the table. How many seconds in 2 minutes? How many degrees in 3 signs? How many signs in 2 degrees? In 3 degrees?

Time.—How is time naturally divided? Recite the table. How many hours in 3 days? How many days in 6 weeks? How many weeks in 5 months? How many months in 8 years? In 11 months how many weeks? In 48 weeks how many months? In 42 days how many weeks? In 56 days? How many months in a year? Name them, and the number of days in each month? How many days has leap-year?

Paper.—Recite the table. How many sheets in 3 quires?

Particulars.—Recite the table. How many in 5 dozen?

REDUCTION.

1. REDUCTION is of two kinds, termed Simple and Compound, by which we can change or reduce a sum or quantity of one kind or denomination, to another, whether it be greater or less, still retaining the same value.

2. When the sum consists of only *one* denomination to be reduced to another, it is called *Simple Reduction*.

3. The operations are all performed either by *Multiplication* or *Division*, for this reason—when a large denomina-

tion is to be reduced to a less, as pounds to drams, or dollars to mills, the operation is by Multiplication, and is called *descending;* but when drams are to be reduced to pounds, or mills to dollars, the operation is by Division, and called *ascending.*

4. Thus, to reduce 2 pounds avoirdupois to drams, multiply the pounds by the number of ounces in a pound, and this will reduce it to ounces; then by the number of drams in an ounce, and this will reduce it to drams: 2lbs. × 16oz. =32oz.; 32×16=512 drams in 2 pounds.

5. Then, to reduce the 512 drams to pounds, divide by 16, and it will give the ounces; then by 16 again, and it will give the pounds: thus, 512÷16=32 ounces; 32÷16=2 pounds.

6. By the above process, sums in Reduction will reciprocally prove each other; and it should be generally practised.

SIMPLE REDUCTION.

RULE.

1. Multiply the sum or quantity by that number of the next lower denomination which it requires to make *one* of its own.

2. If there be one or more denominations between that and that to which it is to be reduced, first reduce it to the next lower than its own, and then to the next lower, and so continue.

3. When low denominations are to be brought to higher, as drams to pounds, cents to dollars, inches to miles, &c., divide by as many of the lower as make one of the higher, and set down what remains (if any) on the right, and so continue.

EXAMPLE.

Reduce 2 cwt. to drams.

(6)

1.   2 cwt.
        4 × qrs. = 1 cwt.
        —
        8 qrs. = 2 cwt.
        28 × lbs. = 1 qr.
        —
        224 lbs. = 8 qrs.
        16 × oz. = 1 lb.
        ——
        1344
        224
        ——
        3584 oz. = 224 lbs.
        16 × drams = 1 oz.
        ——
        21504
        3584
        ——
dr. 16)57344 drams = 3584 oz.
        ——
oz. 16)3584 ounces.
        ——
lbs. 28)224 pounds.
        ——
quarters 4)8 quarters.
        —
        2 cwt.—proof.

Explanation.— First multiply the 2 cwt. by 4, and this will reduce it to quarters, because 4 quarters make 1 cwt.; then multiply the quarter by 28, and this will reduce it to pounds, because 28 pounds make 1 quarter; then multiply by 16, which will reduce it to ounces, because 16 ounces make 1 pound; then multiply the ounces by 16, and it will reduce it to drams, because 16 drams make 1 ounce. Then, to prove it, divide the sum of the drams by the same denominations by which you multiplied, and this will reduce it to 2 cwt.

——— Descending. ——— Ascending. ———

2. Reduce 1 year to seconds.   Thus :—
        1 year = 365 days.
              4
            ——
            1460   4 × 6 =
            6   24 hours.
            ——   = 1 day.
            8760
            60
            ————
            525600 minutes.
            60 sec. = 1 min.
sec. 6,0)3153600,0 seconds. *Ans.*
min. 6,0)52560,0 minutes.
            6)8760 hours.
4 × 6 × 24 4)1460
            ——
            365 days = 1 year—*proof.*

REVIEW.

Reduction.—1. What do you understand by Reduction?
2. When the sum consists of only one denomination? 3.
How are all the operations to be performed? 4. How will
you reduce 2 pounds to drams? 5. How will you reduce
drams to pounds? 6. Proof. *Rule.* 1. How do you multiply? 2. If there be one or more denominations? 3.
When low denominations are to be reduced to higher?

AVOIRDUPOIS WEIGHT.

(1.) Bring 80 pounds to ounces.	*Ans.* 1280.
(2.) Bring 3 cwt. to ounces.	5376.
(3.) Bring 5 cwt. to ounces.	8960.
(4.) Bring 6 cwt. to drams.	172032.
(5.) Bring 1280 ounces to pounds.	80.
(6.) Bring 960 drams to pounds.	3 lbs., 12 oz.
(7.) Bring 3 quarters to drams.	21504.
(8.) Bring 2 tons to drams.	

TROY WEIGHT.

(9.) Bring 12 pounds to ounces.	*Ans.* 144.
(10.) Bring 15 pounds to grains.	86400.
(11.) Bring 20 pounds to grains.	115200.
(12.) Bring 738 ounces to pounds.	61 lbs., 6 oz.
(13.) Bring 1296 grains to ounces.	

APOTHECARIES' WEIGHT.

(14.) Bring 144 ounces to drams.	*Ans.* 1152.
(15.) Bring 8 pounds to grains.	46080.
(16.) Bring 14 pounds to ounces.	168.
(17.) Bring 64 pounds to scruples.	18432.
(18.) Bring 1600 scruples to pounds.	

CLOTH MEASURE.

(19.) Bring 6 yards to quarters.	*Ans.* 24.
(20.) Bring 5 yards to nails.	80.
(21.) Bring 64 quarters to nails,	256.
(22.) Bring 172 quarters to nails.	688.
(23.) Bring 288 nails to yards.	

LIQUID MEASURE.

(24.) Bring 8 gallons to pints.	*Ans.* 64.
(25.) Bring 6 hogsheads to gallons.	378.
(26.) Bring 2 hogsheads to gills.	4032.

(27.) Bring 2 barrels to pints. *Ans.* 504.
(28.) Bring 6794 gills to gallons. 212 galls., 1 qt., 2 gills.
(29.) Bring 872 pints to barrels. 3 bbls., 14 galls., 2 qts.
(30.) Bring 4984 pints to hogsheads.

DRY MEASURE.

(31.) Bring 12 bushels to pecks. *Ans.* 48.
(32.) Bring 832 quarts to bushels. 26.
(33.) Bring 2304 pints to bushels. 36.
(34.) Bring 75 bushels to quarts. 2400.
(35.) Bring 2688 quarts to bushels.

LONG MEASURE.

(36.) Bring 30 yards to feet. *Ans.* 90.
(37.) Bring 60 poles to feet. 990.
(38.) Bring 80 furlongs to poles. 3200.
(39.) Bring 1 mile to inches.
(40.) Bring 4800 poles to furlongs. 120.

LAND OR SQUARE MEASURE.

(41.) Bring 15 roods to poles. *Ans.* 600.
(42.) Bring 20 acres to poles. 3200.
(43.) Bring 90 yards to feet. 810.
(44.) Bring 40 feet to inches. 5760.
(45.) Bring 7200 poles to acres.

SOLID OR CUBIC MEASURE.

(46.) Bring 9 feet to inches. *Ans.* 15552.
(47.) Bring 12 feet of round timber to inches. 20736.
(48.) Bring 7984 inches to feet. 4 ft., 1072 in.

MOTION OT CIRCLE MEASURE.

(49.) Bring 40 degrees to minutes. *Ans.* 2400.
(50.) Bring 27 degrees to seconds. 97200.

TIME.

(51.) Bring 10 years to weeks. *Ans.* 520.
(52.) Bring 360 minutes to hours. 6.
(53.) Bring 4 weeks to hours. 672.
(54.) Bring 90 days to hours. 2160.
(55.) Bring 20 days to minutes. 28800.
(56.) Bring 18 hours to seconds. 64800.
(57.) Bring 20 years to days. 7300.
(58.) Bring 9 days to seconds.
(59.) Bring 7898600 seconds to days.
(60.) Bring 648 months to years.

COMPOUND ADDITION.

By Compound Addition we are taught to add numbers of different denominations, such as pounds, ounces, &c., for the purpose of finding the sum, or amount.

RULE.

1. Write down the numbers so that each denomination may stand directly under each other of its kind, leaving a small space between them.

2. Add the right-hand denomination, the same as in Simple Addition.

3. Then divide that amount by as many as it requires, of that denomination, to make *one* in the next.

4. Set down the remainder under that denomination, and carry the quotient to the next denomination, and add it in: if there be no remainder, set down a cipher.

5. Continue in this way through to the last denomination, which add, the same as in Simple Addition, and set down the whole amount.

Proof, the same as in Simple Addition.

REVIEW.

What is Compound Addition? *Rule.*—1. How do you write down numbers? 2. How do you add? 3. How do you divide? 4. What is done with the remainder? 5. What then? Proof.

EXAMPLES.

	20	4	28	16	16
	T.	cwt.	qr.	lbs.	oz.	dr.
	2	3	2	11	6	8
	4	1	3	16	11	12
	2	4	0	18	10	14
sum	8	9	2	18	13	2

Explanation.—This example belongs to Avoirdupois weight. The first denomination at the right is drams, which add, the same as in Simple Addition, and the amount is 34; divide by 16, because 16 drams make 1 ounce, and we have 2 ounces in the quotient, and 2 drams over; set down the 2 drams, and carry the 2 ounces to the denomination of that name, which add in, and it will make 29 ounces; divide the 29 ounces by 16 for the pounds, and it will give 1 pound and 13 ounces over; set down the 13 ounces, and

6*

COMPOUND ADDITION.

carry the 1 pound to the next denomination; then add that denomination, and it will give 46 pounds; divide the pounds by 28 to get the quarters, and you have 1 quarter and 18 pounds; set down the 18 pounds, and carry the 1 quarter to the denomination of that name, which add in, and you have 6 quarters; which divide by 4, because 4 quarters make 1 cwt., and you have 1 cwt. and 2 quarters; set down the 2 quarters, and carry the 1 cwt. to the next denomination, and add it in; and there will be 9 cwt. and none to carry, because it is less than a ton; then add and set down *all* the next denomination. This is the sum.

AVOIRDUPOIS WEIGHT.

(1.)

T.	cwt.	qrs.	lbs.
2	13	2	12
1	16	3	8
4	11	2	0
1	00	1	8
10	2	1	0

(2.)

cwt.	qrs.	lbs.	oz.
2	3	12	9
4	1	16	12
2	2	14	14
3	0	12	10
5	2	8	6

CLOTH MEASURE.

(7.)

yds.	qrs.	na.
42	3	1
70	1	3
24	2	2
12	1	2
150	1	0

(8.)

E. E.	qrs.	na.
80	4	2
41	3	1
24	2	2
12	2	0
18	2	3

TROY WEIGHT.

(3.)

lbs.	oz.	pwt.
27	8	14
11	6	12
12	7	16
10	8	12
62	7	14

(4.)

lbs.	oz.	pwt.	gr.
21	8	14	16
20	7	9	12
24	8	12	19
17	6	14	18
12	9	8	4

LIQUID MEASURE.

(9.)

galls.	qts.	pts.
21	1	1
18	3	0
12	3	1
19	2	0
72	2	0

(10.)

hhds.	galls.	qts.
2	18	2
12	19	3
74	18	1
60	9	3
18	41	2

APOTHECARIES' WEIGHT.

(5.)

℔.	℥.	ʒ.	℈.
5	8	4	1
19	7	6	2
15	9	7	0
12	8	4	1
53	10	6	1

(6.)

℔.	℥.	ʒ.	℈.	gr.
22	5	4	1	14
21	4	3	2	12
34	2	4	0	14
20	8	2	1	10
14	7	8	1	14

DRY MEASURE.

(11.)

bush.	pks.	qts.
12	3	2
18	2	7
20	1	4
19	2	3
71	2	0

(12.)

bush.	pks.	qts.	pts.
26	2	7	1
18	3	5	0
17	2	0	1
12	0	4	1
16	1	2	1

LONG MEASURE.

(13.)			(14.)			
yds.	ft.	in.	L.	m.	fur.	po.
4	2	7	8	2	6	14
5	2	8	7	2	4	12
10	0	11	20	1	3	16
14	1	5	40	0	2	18
			12	2	1	5
35	1	7				

SOLID OR CUBIC MEASURE.

(17.)		(18.)		(19.)	
T.	ft. 40.	cord.	ft.128.	ft.	in.1728.
47	18	5	18	12	1400
21	19	3	29	18	740
18	30	7	110	19	984
17	24	11	84	27	631
		12	72	40	214
105	11				

LAND OR SQUARE MEASURE.

(15.)			(16.)		
yds.	ft.	in.	A.	R.	po.
10	2	24	112	2	18
16	2	95	14	1	19
18	1	47	78	0	38
25	3	74	12	2	21
			7	1	19
70	0	96			

TIME.

(20.)				(21.)			
da.	h.	m.	sec.	w.	da.	h.	m.
5	20	40	18	2	4	12	18
6	14	38	19	1	4	18	40
7	11	14	16	5	6	7	45
12	12	16	18	9	5	17	35
				6	5	19	25
32	10	49	11				

(22.) Purchased 4 hogsheads of sugar, which weighed as follows: 1st, 4 cwt., 2 qrs., 18 lbs.; 2d, 5 cwt., 3 qrs., 21 lbs.; 3d, 6 cwt., 2 qrs., 12 lbs.; 4th, 6 cwt., 3 qrs., 19 lbs.: what is the weight of them all? *Ans.* 24 cwt., 0 qr., 14 lbs.

(23.) A merchant sold 4 pieces of sheeting; the 1st piece contained 33 yds., 2 qrs., 1 na.; 2d, 35 yds., 3 na.; 3d, 40 yds., 3 qrs., 3 na.; 4th, 45 yds., 2 qrs., 1 na.: required the number of yards. *Ans.* 155 yds. 1 qr., 0 na.

(24.) A farmer disposed of 4 bags of grain, which measured as follows: 1st, 4 bush., 3 pks., 4 qts., 2d, 3 bush., 1 pk., 7 qts.; 3d, 3 bush., 2 pks., 5 qts., 1 pt.; 4th, 4 bush., 1 pk., 5 qts.: how much grain did he sell? *Ans.* 16 bush., 1 pk., 5 qts., 1 pt.

(25.) A grocer purchased 4 hogsheads of vinegar, which measured as follows: 1st, 58 galls., 1 qt., 1 pt.; 2d, 59 galls., 2 qts., 0 pt.; 3d, 58 galls., 1 qt., 1 pt.; 4th, 57 galls., 2 qts.: how much in all? *Ans.* 233 galls, 3 qts., 0 pt.

(26.) A farmer has divided his farm into 5 large fields; the 1st contains 32 A., 1 R., 21 po.; 2d, 27 A., 2 R. 18 po.; 3d, 34 A., 1 R., 30 po.; 4th, 35 A., 1 R., 11 po.; 5th, 26 A., 3 R., 28 po.: required the number of acres in the farm. *Ans.* 156 A., 2 R., 28 po.

(27.) There are 4 persons, A., B., C., and D., whose ages are as follows: A., 76 yrs., 9 mo., 3 w. 4 da.; B., 84 years, 5 mo., 5 da.; C., 92 yrs., 6 m., 3 w., 4 da.; D., 102 yrs., 7 mo., 2 w., 4 da., 6 h.: required their united ages.

Ans. 356 yrs., 5 mo., 2 w., 3 da., 6 h.

(28.) Purchased 4 boat-loads of wood; the 1st boat had 27 C., 94 ft.; 2d, 32 C., 111 ft.; 3d, 28 C., 84 ft.; 4th, 25 C., 47 feet: required the quantity of wood purchased.

Ans. 114 C., 80 ft.

(29.) A merchant sold 4 pieces of cloth; the 1st piece contained 25 yds., 2 qrs., 1 na.; 2d, 26 yds., 3 qrs., 2 na.; 3d, 27 yds., 1 na.; 4th, 24 yds., 2 qrs.: required the number of yards in the 4 pieces, and the value thereof at 5 dollars per yard. *Ans.* 104 yds.; value D520.

(30.) 2 T., 3 cwt., 1 qr., 18 lbs., 12 oz.+5 T., 6 cwt., 2 qrs., 18 lbs., 11 oz., 5 dr.+7 T., 6 cwt., 3 qrs., 20 lbs.+5 T., 2 cwt., 1 qr., 12 oz.

Ans. 19 T., 19 cwt., 1 qr., 2 lbs., 3 oz., 5 dr.

(31.) On a journey of 4 days, I rode the 1st day 42 m., 4 fur., 19 po.; 2d day, 41 m., 5 fur., 30 po.; 3d, 43 m., 7 fur., 25 po.; 4th, 45 m., 4 fur., 19 po.: required the distance travelled in 4 days. *Ans.* 173 m., 6 fur., 13 po.

(32.) In 5 piles of cord-wood, the 1st measures 4 C., 60 ft.; 2d, 7 C., 82 ft.; 3d, 9 C., 110 ft.; 4th, 10 C., 45 ft.; 5th, 11 C., 65 ft.: required the quantity of wood in the 5 piles.

Ans. 43C., 106 ft.

COMPOUND SUBTRACTION.

THIS rule is used when numbers of different denominations are given to be subtracted, or a smaller number taken from a greater, of a like denomination, and show their difference, or remainder.

RULE.

1. Write down the larger number, and directly under it the less number, so that the same denominations shall stand under each other.

2. Begin at the right-hand, and subtract the lower from the upper number, if that be the largest, and set down the remainder.

3. But if the lower number is more than the one above it, then subtract from as many as it takes of that denomination to make *one* in the next; take the difference, and add it to the upper number, and set it down.

4. Carry *one*, and add it to the next lower denomination, and continue through the sum in this manner, and in the last denomination subtract the same as in integers.

Proof, the same as in integers, observing to carry as above directed.

What is Compound Subtraction? 1. How do you write down the given numbers? 2. How do you subtract? 3. If the lower number is more than the one above it? 4. How will you carry when you borrow? Proof.

EXAMPLES.

10	20	4	28	16	16
T.	cwt.	qrs.	lbs.	oz.	dr.
9	7	2	14	11	12
—4	4	1	6	15	4
5	3	1	7	12	8
9	7	2	14	11	12 proof.

Explanation.—Write the value of the respective denominations over each; then, 4 from 12 and 8 remain; 15 from 16 and 1 remains, which added to 11 makes 12; then 1 to carry to 6 is 7, 7 from 14, 7; 1 from 2, 1 ; 4 from 7, 3 ; 4 from 9, 5 : then to prove it, add the two lower numbers, or the *subtrahend* and *remainder*, and their sum will be equal to the upper number, or *minuend*.

AVOIRDUPOIS WEIGHT.		APOTHECARIES' WEIGHT.	

(1.)		(2.)		(5.)				(6.)			
T. cwt. qrs. lbs.		T. cwt. qrs. lbs. oz.		℔.	℥.	ℨ.	Ɔ.	℔.	℥.	ℨ.	Ɔ.
40 12 2 14		25 12 0 14 15		50	7	5	1	90	5	7	2
—30 4 3 22		22 18 1 5 11		24	5	3	2	45	6	5	1
10 7 2 20				26	2	1	2				

TROY WEIGHT.		CLOTH MEASURE.	

(3.)		(4.)		(7.)			(8.)		
lbs. oz. pwt. gr.		lbs. oz. pwt. gr.		yds. qrs. na.			yds. qrs. na.		
47 8 5 14		17 8 12 16		649 2 1			78 0 2		
40 2 4 16		3 7 14 9		552 3 0			47 1 0		
7 6 0 22				96 3 1					

LIQUID MEASURE.

(9.)					(10.)			
T.	hhd.	gals.	qts.	pts	hhds.	gals.	qts.	pts
2	1	40	1	0	41	16	2	1
1	0	15	2	1	12	17	3	0
1	1	24	2	1				

DRY MEASURE.

(11.)				(12.)			
bush.	pks.	qts.	pts.	bush.	pks.	qts.	pts.
450	3	5	1	334	5	6	0
418	2	7	0	290	2	4	1
32	0	6	1				

LONG MEASURE.

(13.)				(14.)				
L.	m.	fur.	po.	fur.	po.	yds.	ft.	in.
70	2	7	20	16	4	4	2	7
62	1	5	24	15	3	1	0	8
8	1	1	36					

LAND OR SQUARE MEASURE.

(15.)			(16.)		
A.	R.	po.	A.	R.	po.
790	3	20	1092	1	28
472	2	25	978	2	30
318	0	35			

SOLID OR CUBIC MEASURE.

(17.)		(18.)		(19.)		
T.	ft. b.	C.	ft.	T.	ft.	in.
240	20	48	112	25	14	260
176	25	31	114	14	16	198
63	45					

MOTION OF CIRCLE MEASURE.

(20.)				(21.)			
sig.	°.	'.	''.	sig.	°.	'.	''.
20	7	40	18	47	11	20	50
18	5	15	20	20	8	17	55
2	2	24	58				

TIME.

(22.)					(23.)			
Y.	m.	w.	da.	h.	Y.	m.	w.	da.
47	8	2	5	14	90	4	2	5
25	5	3	4	11	40	7	3	6
22	2	3	1	3				

(24.)				(25.)				
da.	h.	m.	sec.	m.	w.	da.	h.	m.
40	18	50	20	74	2	5	18	40
17	12	55	18	69	0	6	14	55
23	5	55	2					

(26.) A grocer purchased a cask of sugar weighing 9 cwt., 1 qr., 18 lbs., 11 oz.; if he should sell 4 cwt., 12 lbs., 8 oz., how much would remain?

(27.) A merchant had a piece of sheeting containing 33 yards, 1 qr., 2 na.: he sold to W. 9 yds., 1 qr., 1 na.; to M. 11 yds., 2 qrs., 1 na.: how many yards has he remaining of the piece?   *Ans.* 12 yds., 2 qrs., 0 na.

(28.) A farmer has 2 farms; the 1st contains 274 A., 1 R., 20 po.; 2d, 304 A., 1 R., 10 po.: he wishes to give his only son one half, and employed a surveyor to measure off 289 A.,

1 R., 15 po., supposing this quantity would be an equal division of all his land; was he correct in his calculation?

(29.) A grocer purchased 2 hhds. of vinegar, and sold 84 galls., 2 qts., 1 pt.; what quantity remained?

Ans. 41 galls., 1 qt., 1 pt.

(30.) Take 454 bush., 2 pks., 1 qt., from 620 bush., 1 pk., 2 qts.   *Ans.* 165 bush., 3 qts., 1 pt.

(31.) If it be 264 m., 4 fur., 19 po., to Boston, and 194 m., 7 fur., 35 po., to Albany, how much farther is it to Boston than to Albany?   *Ans.* 69 m., 4 fur., 24 po.

(32.) From 94 yrs. take 33 yrs., 4 mo., 2 w., 5 da.

Ans. 60 yrs., 7 mo., 1 w., 2 da.

(33.) From a pile of wood, containing 42 cords, was sold 21 cords and 96 cubic feet; what was the quantity of wood left?   *Ans.* 20 C., 32 ft.

(34.) 21 yrs., 2 mo., 3 w., 4 da., 5 h.—12 yrs., 5 mo., 3 w., 4 da., 11 h.   *Ans.* 8 y., 8 mo., 3 w., 6 da., 18 h.

(35.) 9 sig., 11°, 22′, 25″—5 sig., 15°, 35′, 42″.

Ans. 3 sig. 25°, 46′, 43″.

(36.) 4 mo., 1 w., 3 da., 1 h., 6 m.—2 mo., 1 w., 3 da., 2 h., 20 m.   *Ans.* 1 mo., 3 w., 6 da., 22 h., 46 m.

(37.) 1598 bush., 3 pks., 7 qts., 0 pt.—995 bush., 3 pks., 5 qts., 1 pt.   *Ans.* 603 bush., 0 pk., 1 qt., 1 pt.

(38.) 649 hhds., 12 galls., 2 qts., 1 pt.—281 hhds., 14 galls., 3 qts., 0 pt.   *Ans.* 367 hhds., 60 galls., 3 qts., 1 pt.

(To find the difference between two given dates.)

Rule.—Write down the larger number, and under it the less number. If the number of days in the less number be more than in the greater, subtract from as many as there be days in the month mentioned in the less number, and add in the days in the greater number, which set down, and carry one to the month: then subtract in the usual way.

(39.) A gentleman was born March 17, 1796; required his age on June 10, 1847.

$$
\begin{array}{ccc}
1847 & 6 & 10 \\
1796 & 3 & 17 \\
\hline
51 & 2 & 24 \\
\end{array}
$$

(40.) The battle of Lexington occurred April 19, 1775; Presidents Adams and Jefferson died July 4, 1826: required the time that elapsed between those two periods.

Ans. 51 yrs., 2 mo., 15 da.

(41.) General William Hull surrendered his army, at Detroit, August 16, 1812; General Jackson gained the victory of New Orleans, January 8, 1815; required the lapse of time between those two events.

COMPOUND MULTIPLICATION.

THE use of the following rule is to perform a number of additions of different denominations, or when numbers of this kind are to be multiplied.

(When the number does not exceed twelve.)

RULE.

1. Write down the multiplicand, and write the quantity of the several denominations over each, as directed in Subtraction.
2. Then write the multiplier under the lowest denomination, at the right-hand.
3. Multiply that denomination, and divide it by as many as it takes of that denomination to make *one* in the next, and set down the remainder, carry the quotient to the product of the next denomination, and so continue.

Proof, as in multiplication of integers, or Division, as most correct.

EXAMPLES.

10 T.	20 cwt.	4 qrs.	28 lbs.	16 oz.	16 dr.	
4	9	2	17	12	13	7×
31	7	2	12	9	11	

Explanation.— First 7×13 =91÷16=5 and 11 over; now 12×7=84 and 5 are 89÷16 =5 and 9 over, set down the 9; then 17×7=119 and 5 are 124×28=4 and 12 over, set down the 12; 2×7=14 and 4 are 18÷4 are 4 cwt., and 2 qrs. over, set down the 2 qrs.; now 9×7=63, and the 4 cwt. make 67÷20=3 tons and 7 cwt. over, set down the 7 cwt.; now 4×7=28, and the 3 tons are 31 tons.

(1.)   lbs. oz. pwt. gr.
(Troy)   8   4   5   3
                   3×
     25   0   15   9

(2.)   L.   m.   for.   po.
     14   2   5   12
                  5×
     74   1   2   20

(3.)   A.   R.   po.
     9   3   20
             6×
     59   1   0

(4.)   bush. pks. qts. pts.
     12   3   4   1
                  7×
     90   0   7   1

(5.) 12 lbs., 4 oz., 15 dwt., 14 gr. ×5=
       Ans. 61 lbs., 11 oz., 17 dwt., 22 grs.

(6.) 11 T., 2 cwt. 3 qrs., 11 lbs., 6 oz. ×7=
       Ans. 77 T., 19 cwt., 3 qrs., 23 lbs., 10 oz.

(7.) 5 T., 12 cwt., 2 qrs., 12 lbs., 6 oz., 11 drs. ×8=
       Ans. 45 T., 0 cwt., 3 qrs., 15 lbs., 5 oz., 8 drs.

(8.) 12 L., 1 m., 6 fur., 20 po., 0 yds., 1 ft., 4 in. ×9=
       Ans. 113 L., 1 m., 2 fur., 20 po., 4 yds., 0 ft., 0 in.

(9.) 20 C., 112 ft., 589 in. ×5=
       Ans. 104 C., 49 ft., 1217 in.

(10.) 140 yds., 3 qrs., 2 na. ×11=
       Ans. 1549 yds, 2 qrs., 2 na.

(11.) 12 T., 1 hhd., 5 galls., 3 qts., 1 pt. ×12=
       Ans. 147 T., 1 hhd., 7 galls., 2 qts., 0 pt.

(12.) 14 yrs., 11 mo., 2 w., 4 da., 6 h., 5 m. ×9=
       134 yrs., 8 mo., 3 w., 3 da., 6 h., 45 m.

(13.) 4 sig., 20°, 40′, 25″ ×7=
       Ans. 32 sig., 24°, 42′, 55″.

(14.) 20 A., 1 R., 12 po., 9 ft. ×8=
       Ans. 162 A., 2 R., 20 po., 6 ft.

REVIEW.

What is the use of Compound Multiplication? 1. How will you begin to write the denominations? 2. Where will you write the multiplier? 3. How will you multiply? Proof.

(When the multiplier exceeds twelve, and is the product of any two figures in the multiplication table.

RULE.

Multiply the sum by one of the figures, and that product by the other, and the last product will be the answer.

7

(15.) Multiply 20 A., 1 R., 15 po., by 25.   5×5=25.

$$
\begin{array}{ccc}
20 & 1 & 15 \\
 & & 5\times \\
\hline
101 & 2 & 35 \\
 & & 5\times \\
\hline
\end{array}
$$

Ans. 508   2   15

(16.) 7 A., 1 R., 19 po. ×16.   *Ans.* 117 A., 3 R., 24 po.

(17.) 25 A., 12 po. ×20.   *Ans.* 501 A.,   2 R.

(18.) 90 m., 7 fur., 18 po. ×54.

Ans. 4910 m., 2 fur., 12 po.

(19.) 5 T., 12 cwt., 1 qr., 12 lbs., 6 oz. ×36.

Ans. 202 T., 4 cwt., 3 qr., 5 lb., 8 oz.

(20.) 15 cwt., 3 qrs., 11 lbs., 12 oz., 9 drs. ×42.

Ans. 665 cwt., 3 qrs., 18 lbs., 15 oz., 10 drs.

(21.) 159 yds., 1 qr., 3 na. ×44.

Ans. 7015 yds., 1 qr., 0 na.

(22.) 16 C., 82 ft., 350 in. ×48.

Ans. 798 C. 105 ft., 1248 in.

(23.) 414 bush., 2 pk., 5 qts., 1 pt. ×64.

Ans. 26539 bush., 0 pk., 0 qt., 0 pt.

(24.) 15 yrs., 4 mo., 1 w., 5 da., 11 h., 20m. ×72.

Ans. 1106 yrs., 7 mo., 1 w., 2 da., 0 h., 0 m.

(25.) 19 yds., 2 qrs., 3 na. ×108.

Ans. 2126 yds., 1 qr., 0 na.

(26.) 11 mo., 2 w., 4 da., 6 h., 20 m., 15 sec. ×77.

Ans. 897 mo., 0 w., 6 da.. 7 h., 59 m., 15 sec.

(When the multiplier consists of three or more figures, and is not a composite number.)

RULE.

1. Multiply the simple number by each of the denominations separately, and reduce each product to the highest denomination named. 2. Then add the several products together, and their sum will be the answer sought.

(27.) 4 T., 6 cwt., 2 qrs., 20 lbs. ×141.

Ans. 611 T., 1 cwt., 2 qrs., 20 lbs.

(28.) 150 L., 2 m., 4 fur., 20 po. ×26.

Ans. 3921 L., 2 m., 5 fur., 0 po.

(29.) 4 hhds., 27 galls., 2 qts., 1 pt. ×15.

Ans. 66 hhds., 36 galls., 1 qt., 1 pt.

(30.) 15 lbs., 7 oz., 12 dwt., 19 grs. ×8.

Ans. 125 lbs., 1 oz., 2 dwt., 8 grs.

(31.) How much cloth will it require to make 52 coats, allowing to each 2 yds., 1 qr. 2 na.? *Ans.* 123 yds., 2 qrs., 0 na.

(32.) There are 14 piles of wood, each containing 7 C., 114 ft.; what is the whole quantity? *Ans.* 110 C. 60 ft.

(33.) What is the weight of 20 chests of tea, each weighing 4 cwt., 1 qr., 12 lbs.? *Ans.* 87 cwt., 0 qr., 16 lbs.

REVIEW.

When the multiplier exceeds twelve, what is the rule? When the multiplier consists of three or more figures, what is the rule?

COMPOUND DIVISION.

This rule is the reverse of Compound Multiplication: it shows how often a given quantity is contained in another of different denominations. Multiplication and Division will reciprocally prove each other; and this should be generally practised.

GENERAL RULE.

1. If the divisor does not exceed twelve, proceed as in Short Division of whole numbers.

2. Divide the first denomination on the left-hand by the divisor; multiply the remainder, if any, by the number of the second denomination contained in a unit of the first, and add the second to the product; divide as before.

3. When the divisor is a *composite* number, and is greater than twelve, divide by the factors successively.

4. When the divisor is a *prime* number greater than twelve, divide by the whole divisor as in Long Division. Proof, by Multiplication.

REVIEW.

What is Division the reverse of? What two rules will prove each other? Why will they prove each other? What should be generally practised? 1. When the divisor does not exceed twelve, what will you do? 2. How will you divide the first denomination? 3. When the divisor is a *composite* number, how can you divide? 4. How will you divide when the divisor is a *prime* number? Proof. What is a *prime* number? *Ans.* A prime number is one which can be measured only by itself or a unit, as 3, 7, 23, &c. What is a *perfect* number? *Ans.* A perfect number is equal to the sum of all its aliquot parts.

EXAMPLES.

(1.)
$$\begin{array}{r} \overset{4}{} \quad \overset{28}{} \\ \text{cwt. qrs. lbs.} \\ 5)14 \quad 3 \quad 15 \\ \hline 2 \quad 3 \quad 25+2 \end{array}$$

Explanation.—5 in 14=2 times and 4 over; 4×4+3=19; 5 in 19, 3 times and 4 over; 4×28+15= 127; 5 in 127, 25 times and 2 over.

(2.)
$$\begin{array}{r} \text{cwt. qrs. lbs.} \\ 5)64 \quad 1 \cdot 25 \\ \hline 12 \quad 3 \quad 16+1 \end{array}$$

(3.)
$$\begin{array}{l} \text{yr. mo. w. da. h.} \\ 14)26 \quad 8 \quad 3 \quad 4 \quad 6 \\ \quad 14 \\ \hline \quad 12 \\ \quad 12\times \\ \hline 14)152+8 \\ \quad 14 \\ \hline \quad 12 \\ \quad \underline{4\times} \\ 14)51+3 \\ \quad 42 \\ \hline \quad \,9 \\ \quad \underline{7\times} \\ 14)67+4 \\ \quad 56 \\ \hline \quad 11 \\ \quad \underline{24\times} \\ \quad 44 \\ \quad 22 \\ \hline 14)270+6 \\ \quad 14 \\ \hline \quad 130 \\ \quad 126 \\ \hline \quad \,4 \end{array}$$

yr. m. w. da. h.
(1   10 ·3   4   19+4   *Ans.*
  14×

$$\overline{26 \quad 8 \quad 3 \quad 4 \quad 6} \text{ proof.}$$

Explanation.—14 in 26, once and 12 over: 12×12 mo.+8 mo.=152÷ 14=10 mo. and 12 over; 12×4 w. +3=51÷14=3 w. and 9 over; 9×7 +4=67÷14=4 days and 11 over; 11×24 h.+6=270÷14=19 h. and 4 over. Then, to prove it, multiply the quotient by the divisor, and add in the remainder, and you produce the dividend.

(4.) 9 lbs., 6 oz., 11 dwt.÷4.

Ans. 2 lbs., 4 oz., 12 dwt.+3.

(5.) 11 lbs., 4 oz., 1 dr., 2 scr., 12 grs.÷5.

Ans. 2 lbs., 3 oz., 1 dr., 2 scr., 18 grs.+2.

(6.) 6 cwt. 1 qr., 11 lbs., 4 oz.÷3.

Ans. 3 cwt., 0 qr., 13 lbs., 1 oz.+1.

(7.) 4 T., 6 cwt., 14 lbs. 12 oz., 8 dr.÷4.

Ans. 1 T., 1 cwt., 17 lbs., 11 oz., 2 dr.

(8.) 14 L., 2 m., 4 fur., 8 po.÷5.

Ans. 2 L., 2 m., 7 fur., 9 po.+3.

(9.) 9 L., 1 m., 3 fur., 0 po., 4 yds.÷6.

Ans. 1 L., 1 m., 5 fur., 33 po., 2 yds.+3.

(10.) 25 m., 5 fur., 25 po., 2 yds., 2 ft.÷7.

Ans. 3m., 5 fur., 15 po., 0 yd., 1 ft.+1.

(11.) 5 fur., 17 po., 3 yds., 2 ft., 8 in., 2 bar.-c.÷8.

Ans. 27 po., 1 yd., 0 ft., 6 in., 1 bar.-c.

(12.) 27 A., 1 R., 9 po., 12 ft.÷9.

Ans. 3 A., 0 R., 5 po., 8 ft.+6.

(13.) 70 C., 94 ft., 1294 in.÷10.

Ans. 7 C., 9 ft., 820 in.+6.

(14.) 27 yds., 2 qrs., 3 na.÷11.

Ans. 2 yds., 2 qrs., 0 na.+3.

(15.) 45 hhds., 25 galls., 3 qts., 1 pt.÷12.

Ans. 3 hhds., 49 galls., 1 qt., 1 pt.+3.

(16.) 119 bush., 2 pks., 4 qts.÷8.

Ans. 14 bush., 3 pks, 6 qts., 1 pt.

(17.) 25 yds., 1 qr., 3 na.÷7.

Ans. 3 yds., 2 qrs., 2 na.+1.

(18.) 42 yrs., 6 mo., 3 w., 8 da., 9 h.÷9.

Ans. 4 yrs., 8 mo., 3 w., 0 da., 22 h.+3.

(19.) 41 sig., 4°, 17′, 18″÷8. Ans. 5 sig., 4°, 17′, 9″+6.
(20.) 480 cwt., 3 qrs.÷28. Ans. 17 cwt., 0 qr., 19 lbs.
(21.) 50 T., 12 cwt., 3 qrs., 8 lbs.÷45.

Ans. 1 T., 2 cwt., 2 qrs., 0 lb.+36.

(22.) 125 cwt., 2 qrs., 18 lbs.÷72.

Ans. 1 cwt., 2 qrs., 27 lbs.+34.

(23.) 478 yds., 1 qr., 2 na.÷39.

Ans. 12 yds., 1 qr., 0 na.+10.

(24.) 587 m., 2 fur., 7 po.÷51.

Ans. 11 m., 4 fur., 4 po.+43.

(25.) 678 cwt., 2 qrs., 12 lbs.÷125.

Ans. 5 cwt., 1 qr., 20 lbs.+4.

(26.) Divide 225 bush., 2 pks., 6 qts., among 4 persons.

Ans. 56 bush., 1 pk., 5 qts.+2.

7*

(27.) Divide 576 A., 2 R., 18 po., among 7 men.

                            Ans. 82 A., 1 R., 14 po.

(28.) Divide a hogshead of wine equally among 20 persons.   *Ans.* 3 galls., 0 qt., 1 pt.+4=1 gi. nearly.

(29.) If it require 3 yds. to make a coat, how many coats can be made from 72 yds., of broadcloth?

                                Ans. 24 coats.

(30.) What will be the share of 1 man, if 250 T., 12 cwt., 28 lbs., be equally divided among 75 men?

                      Ans. 3 T., 6 cwt., 3 qrs., 8 lbs.+72.

(31.) What time will you require to walk 500 m., at the rate of 25 m. a-day?   *Ans.* 20 da.

(32.) If 20 loads of hay contain 42 T., 5 cwt., what is the weight of each load?   *Ans.* 2 T., 2 cwt., 1 qr.

MISCELLANEOUS QUESTIONS IN THE COMPOUND RULES.

(1.) If a man can travel in one day 20 m., 4 fur.; another day 25 m., 3 fur., 18 po.; another day 22 m., 4 fur., 36 po., required the distance.   *Ans.* 71 m., 1 fur., 14 po.

(2.) In 3 tracts of Michigan land, the 1st contains 240 A., 2 R., 5 po.; 2d, 268 A., 3 R., 25 po.; 3d, 468 A., 2 R., 20 po.; required the number of acres in the 3 tracts.

                        Ans. 978 A., 0 R., 10 po.

(3.) Add 2 cwt., 1 qr., 25 lbs.; 4 cwt., 1 qr., 16 lbs.; 6 cwt., 2 qrs., 8 lbs.; 21 cwt., 3 qrs., 20 lbs.; 15 cwt., 2 qrs., 18 lbs., together.   *Ans.* 51 cwt., 0 qr., 3 lbs.

(4.) Bought 3 casks of vinegar; the 1st contained 18 galls., 2 qts., 1 pt., 2 gi.; 2d, 21 galls., 2 qts., 0 pt., 3 gi.; 3d, 24 galls.; 3 qts., 1 pt., 2 gi.: required the quantity in the 3 casks.   *Ans.* 65 galls., 0 qts., 1 pt., 3 gi.

(5.) Bought a piece of broadcloth containing 42 yds., 2 qrs., 3 na.; and sold 2 suits from it, each containing 5 yds., 1 qr., 2 na.: how much have I remaining of the piece?

                        Ans. 31 yds., 3 qrs., 3 na.

(6.) If I should sell A. 6 galls., 1 qt., 1 pt.; B. 7 galls., 2 qts., 0 pt., of molasses, from a cask containing 93 galls., 2 qts., 1 pt., how much would remain?

                        Ans. 79 galls., 3 qts., 1 pt.

(7.) From 70 bush. take 54 bush., 1 pk., 3 qts.

                        Ans. 15 bush., 2 pks., 5 qts.

(8.) From 8 sq. yds. take 5 ft., 96 in.
Ans. 7 yds., 3 ft., 48 in.
(9.) From 2 lbs., troy, take 8 oz., 12 dwt., 18 grs.
Ans. 1 lb., 3 oz., 7 dwt., 6 grs.
(10.) From 2 galls. take 1 qt., 1 pt., 1 gi.
Ans. 1 gall., 2 qt., 0 pt., 3 gi.
(11.) Required the weight of 4 chests of tea, each weighing 2 cwt., 1 qr., 12 lbs.   *Ans.* 9 cwt., 1 qr., 20 lbs.
(12.) Required the weight of 5 hhds. of sugar, each weighing 5 cwt., 1 qr., 14 lbs. ●   *Ans.* 26 cwt., 3 qrs., 14 lbs.
(13.) In 14 pieces of sheeting, each containing 31 yds., 2
qrs., 1 na., how many yards in all?
Ans. 441 yds., 3 qrs., 2 na.
(14.) A farmer has 5 bins of wheat, containing 47 bush.,
1 pk., 4 qts., 1 pt. each; how much in all?
Ans. 236 bush., 3 pks., 6 qts., 1 pt.
(15.) From 1 yr., 6 mo. take 9 m., 2 w., 2 da.
Ans. 8 mo., 1 w., 5 da.
(16.) Bought 3 loads of cord-wood, the 1st contained 1 C.,
42 ft.; 2d, 1 C., 92 ft.; 3d, 1 C., 86 ft.: how much in all?
Ans. 4 C., 92 ft.
(17.) From 7 galls., take 1 gall., 1 qt., 1 pt., 1 gi.
Ans. 5 galls., 2 qts., 0 pt., 3 gi.
(18.) From 4 A., 2 R., 6 po., take 36 po.
Ans. 4 A., 1 R., 10 po.
(19.) Multiply 2 yds., 2 qrs., 2 na., by 61.
Ans. 160 yds., 0 qr., 2 na.
(20.) 740 A., 3 R., 4 po.+672 A., 0 R., 16 po.
Ans. 1412 A., 3 R., 20 po.

COMPOUND REDUCTION.

IN Simple Reduction but one denomination is given;
whereas in Compound Reduction, several denominations are
given to be reduced to one denomination, still retaining the
same value: as, 1 pound, 5 ounces, avoirdupois, would be
21 ounces, because in 1 pound there are 16 ounces+5
ounces=21 ounces: 21 oz.=1 lb., 5 oz.

RULE.

1. Multiply the highest denomination by as many of the
next less as it requires to make *one* of that, and add in the

second denomination, if any; then multiply by the next less denomination, and add in, if any; continue, in this way through the sum.

2. When small denominations are to be reduced to larger denominations, reverse this operation, and begin by dividing in the same manner as you are directed above to multiply: by this reversion of the operation the proof is obtained. (*See examples.*)

(1.) Reduce 5 cwt., 2 qrs., 12 lbs., 11 oz., 10 dr., to drams.

cwt. 5
4×
—
20
2+
—
22 qrs.
28×
—
176
44
—
616
12+
—
628 lbs.
16×
—
3768
628
—
10048
11+
—
10059 oz.
16×
—
60354
10059
10+
—
160954 drs. *Ans.*

Proof, 16)160954 drams.
—
16)10059+10 dr.
—
28)628+11 oz.
—
4)22+12 lbs.
—
cwt. 5+ 2 qrs.

5 cwt., 2 qrs., 12 lbs., 11 oz., 10 dr.

Explanation.—Multiply the 5 cwt. by 4 qrs., and add in the 2 quarters, and you have 22 quarters; then multiply 22 quarters by 28 pounds, and add in the 12 pounds, and you have 628 pounds; then multiply 628 pounds by 16 ounces, add in the 11 ounces, and you have 10059 ounces; then multiply the ounces by 16 drams, add in 10 drams, and you have 160954, which is the answer.

To prove the work, you can divide by Short or Long Compound Division.

REVIEW.

What is given in Simple Reduction? What do you understand by Compound Reduction? *Rule.*—1. How do you multiply by this rule? 2. How can you reduce small denominations to larger, retaining the same value? In the 1st example, why do you multiply by 4? Why do you multiply by 28? Why do you multiply by 16? Why do you multiply by 16 a second time? When you multiplied, why did you add those denominations to the product? Explain the method of proof.

EXAMPLES.

(2.) Reduce 1 cwt., 1 qr., 7 lbs., to pounds. *Ans.* 147 lbs.

(3.) Reduce 2 cwt., 3 qrs., 12 lbs., 6 oz., to ounces.
Ans. 5126.

(4.) Reduce 1 T., 4 cwt., 18 lbs., 12 oz., to drams.
Ans. 692928.

(5.) Reduce 2 T., 5 cwt., 12 lbs., 4 oz., 6 dr., to drams.
Ans. 1293382.

(6.) Reduce 12 bush., 3 pks., to pecks. 51.

(7.) Reduce 14 bush., 2 pks., to quarts. 464.

(8.) Reduce 8 mo., 1 w., 4 da., to days. 235.

(9.) Reduce 12 yds., 2 ft., 4 in., to inches. 460.

(10.) Reduce 2 L., 1 m., 4 fur., to poles. 2400.

(11.) Reduce 4 A., 2 R., 4 po., to poles. 724.

(12.) Reduce 21 hhds., 12 galls., 1 qt., to quarts. 5341.

(13.) Reduce 17 lbs., 11 oz., 4 dr., 2 scr., to scruples. 5175.

(14.) Reduce 27 yds., 2 qrs., 2 na., to nails. 442.

(15.) Reduce 40 C., 97 ft., 159 in., to solid inches.
Ans. 9015135.

(16.) Reduce 108 lbs., 9 oz., 18 dwt., to grains (troy).
Ans. 626832.

(17.) Reduce 9 m., 3 fur., to barley-corns. 1782000.

(18.) Reduce 128 A., 3 R., 18 po., to poles. 20611.

(19.) In 1286 acres how many perches? 205760.

(20.) In 908 m., 4 fur., how many inches? 57562560.

(21.) In 94 hhds. of wine how many pints? 47376.

(22.) In 12 tons of round timber how many inches? 829440.

(23.) In 4 sig., 3°, 17', 18", how many seconds? 443838.

(24.) In 2016 nails how many yards? 126.

(25.) In 87840 poles how many acres? 549.

(26.) In 78916 drams avoirdupois how many pounds?
Ans. 308 lbs., 4 oz., 4 dr.

(27.) In 276848 pints how many hogsheads?
Ans. 549 hhds., 19 gall.

(28.) In 678947 inches how many poles?

Ans. 3428 po., 5 yds., 1 ft., 11 in.

(29.) In 648248 gills how many hogsheads?

Ans. 321 hhds., 34 galls., 3 qts.

(30.) In 78964860 seconds how many days?

Ans. 913 da., 22 h., 41 m.

(31.) In 8790660 minutes how many weeks?

(32.) In 9878 quarts how many bushels?

(33.) In 6897864 cubic inches how many cords?

(34.) In 1089060 seconds how many signs?

(35.) In 789706582 seconds how many years of 365 days?

PROPORTION,
OR SINGLE RULE OF THREE.

1. PROPORTION is a comparative relation of one thing, or number, to another.

2. Proportion, when complete, consists of four numbers or terms, and the product of the first and fourth terms (which are the extremes) is equal to the product of the second and third terms (middle terms or means).

3. Therefore, it is evident that any one of the four terms is readily obtained when the other three are given.

4. As 2 : 4 :: 8 : 16—which should be read thus : as 2 is to 4, so is 8 to 16; that is, 2 bears the same proportion to 4 as 8 does to 16, for 2 is half of 4, and 8 is half of 16.

5. The two middle terms are the *means*, and the first and fourth terms the extremes: $2 : (4 :: 8) : 16$

$$\frac{4 \times}{32} \quad \frac{2 \times}{32}$$

Here we see that the product of the means is equal to the product of the extremes. If 2 pounds of tea cost 4 dollars, what will 8 pounds cost?

6. Here the price of the tea is 2 dollars per pound; and if 2 pounds will cost 4 dollars, it is evident that 8 pounds will cost 16 dollars, because $2 \times 2 = 4 : 2 \times 8 = 16$; and this principle will apply in all cases of proportion, for if 16 dollars will purchase 8 pounds, 4 dollars will purchase 2 pounds. Here the statement is reversed, but the *proportion* is the same, for $16 : (8 :: 4)\ 2$.   The product of the *means* and *extremes* are the same, and this

$$\frac{2}{32} \quad \frac{4}{32}$$

must always be the case in *proportion*, otherwise the terms will not be proportional.

7. The terms may be distinguished in the following manner: the *first* term is preceded by an *if*, or supposition; and the *second* by a demand—"what will," "what cost," &c.

Rule for stating.—1. Place that term in the third place, which is of the same name or kind, with that in which the answer is required; and if the answer requires to be *greater* than the third term, set the greater of the two remaining numbers for the second term, and the other number for the first term.

2. But if the answer requires to be less than the third term, set the *less* of the two remaining numbers in the second place, and the other in the first place.

If 5 lbs. cost D10, what will 15 lbs. cost? Here the 4th term, or answer, requires to be greater than the third term, because 15 lbs. will cost more than 5 lbs. at the same price; therefore the statement will stand thus:—

As 5 lbs. : (15 lbs. :: 10D.) : 30 D.

$$\frac{10 \times}{150} \qquad \frac{5 \times}{150}$$

As 5 lbs. are to 15 lbs., so are 10D. to 30D.

Then, if 30D. will purchase 15 lbs., how many pounds will 10D. buy? Here the answer requires to be *less* than the third term, because 10D. will purchase less than 30D. at the same price; therefore state as follows:—

As 30D. : (10D. :: 15 lbs.) : 5 lbs. As 30D. are to 10D.

$$\frac{5}{150} \qquad \frac{10}{150} \qquad \begin{cases} \text{so are 15 lbs. to 5 lbs.; and} \\ \text{the product of the } means \end{cases}$$

and *extremes* are equal.

Rule for working.—1. Reduce the third term to the lowest denomination mentioned in it.

2. Reduce the first and second terms to the lowest denomination mentioned in either of them.

3. Multiply the second and third terms together, and divide by the first, and the quotient (the fourth term) will be the answer, in the same name to which the third was reduced; then bring this denomination into the answer required.

4. When 1, or unity, is one of the terms, a formal statement is not required, as it can be solved by multiplication or division.

Proof.—1. Multiply the first and fourth terms together (extremes). 2. Multiply the second and third terms together (means). If the four numbers or terms are proportional, their product will be equal. 3. Invert the question, as in the above example.

REVIEW.

1. What is Proportion? Of how many numbers or terms must it consist? Which of the two terms are the extremes? Which of the two terms are the means? 3. How many terms are given? How many terms are required? 4. How will you read the first statement? What proportion is 2 to 4? 8 to 16? 5. What can you say of the product of the means and extremes? 6. What is the price of the tea, and why will 16 dollars purchase but 8 pounds? How will you know when the terms are proportional? How can you distinguish the terms? *Rule for stating.*—1. What term will you write in the third place? 2. If the answer requires to be less than the third term? *Rule for working.* 1. What will you do with the third term? 2. What will you do with the first and second terms? 3. Which of the two terms will you multiply together? Which of the terms is the divisor? Which of the terms will be the answer? What will be the name or denomination of the quotient or fourth term? What must you do with the quotient? 4. What can you do when 1 or unity is one of the terms? Proof—1, 2, and 3.

EXAMPLES AND QUESTIONS.

(1.) If 5 lbs. of sugar cost 65 cts., what will 10 lbs. cost? *Ans.* D1.30.

As 5 lbs. : 10 lbs. :: 65 c.

$$10\times$$
$$5)\overline{6.50}$$
$$D1.30 \; Ans.$$

As 1.30 : 65 :: 10
D.   cts.   lbs.

$$10\times$$
$$1.30)\overline{6.50}$$
5 lbs. proof.

(2.) If 3 lbs. of tea cost D1.25, how much will 7 lbs. cost?

As 3 lbs. : 7 lbs. :: D1.25

$$7\times$$
$$3)\overline{8.75}$$
$$D2.91.6\frac{2}{3}$$

(3.) If 6 lbs. of butter cost 72 cts., how much can you have for D1.44?

As .72 : 1.44 :: 6 lbs.

$$6\times$$
$$72)864(12 \text{ lbs. } Ans.$$
$$\underline{72}$$
$$144$$
$$144$$

(4.) If D50 will buy 20 yds. of cloth, how many yards can you have for D75?

As D50 : 75 :: 20 yds.

$$20\times$$
$$5,0)\overline{150,0}$$
$$30 \text{ yds. } Ans.$$

(5.) If 5 yds. cost D14.50, what will 12 yds. cost?
Ans. D34.80.

(6.) If 9 bushels of wheat cost D15.50, what will 7 bushels cost?   *Ans.* D12.05.5.+

(7.) If 6 bushels of onions cost D9.50, what will 50 bushels cost?   *Ans.* D79.16.6.+

(8.) If 1 lb. of sugar cost 9½ cts., what will 75 lbs. cost?   *Ans.* D7.12½.

(9.) If 6 bushels of corn cost D12.50, what will 45 bushels cost?   *Ans.* D93.75.

(10.) When flour is worth D8.00 per barrel of 196 lbs., what is 1 lb. worth?   *Ans.* cts.4.08.+

(11.) What cost a hogshead of vinegar at 5 cts. per pint?   *Ans.* D25.20.

(12.) What cost 2 hhds. of molasses at 9 cts. per quart?   *Ans.* D45.36.

(13.) If a bushel of cloverseed is worth D5.00, what is 1 pint worth?   *Ans.* cts.7.8½ m.

(14.) If 5 men can do a piece of work in 25 days, how many days will it require for 3 men to do the same? *Ans.* 41⅔.

(15.) If 1 ton of hay cost D16, what cost 1 cwt.?
                                                    Ans. 80 cts.

(16.) If 8 men can harvest a field of grain in 32 days, in what time will 24 men do it?   *Ans.* 10⅔ days.

(17.) How many bushels of wheat at D1.12 can I have for D81.76?   *Ans.* 73.

(18.) How many yards of sheeting may be bought for D38.40, when 8 yds. cost D3.20?   *Ans.* 96.

(19.) If 500 ft. of pine boards cost D7.50, what will 1750 feet cost?   *Ans.* D26.25.

(20.) What is the value of 1 lb. of cast-steel at D22.00 per cwt.?   *Ans.* cts.19.6.+

(21.) If 75 lbs. of tea cost D87.50, how much will 125.5 lbs. cost?

(22.) If you can earn 28.5 cts. in a day, how much is it in a year?

(23.) If 97½ lbs. of beef cost D8.50, how much must you pay for 1500 lbs.?   *Ans.* D130.76.9.+

(24.) If 225 bushels of oats cost D120, what are 65 bushels worth?   *Ans.* D34.66.6.+

(25.) If 280 acres of land are worth D78.90, what are 95 acres worth?   *Ans.* D2676.96.4.+

(26.) How much is 6.75 yds. of cloth worth at 2.50 per yard?   *Ans.* D16.87½.

(27.) If 3.25 yds. of cloth cost D9.50, what will 12.5 yds. cost?   *Ans.* D36.53.8.+

8

(28.) If 1.5 bushels of wheat cost 2.75, what will 25 bushels cost?   *Ans.* D45.83.3.+

(29.) When corn is selling at 75 cts. per bushel, what will 175.5 bushels cost?

(30.) What would 12½ bushels of oats cost at 62½ cts. per bushel?   *Ans.* D7.81¼.

(31.) Bought 3 loads of corn at 87.5 cts. per bushel; one load has 27.5 bushels, and each of the others 25 bushels: required the cost.   *Ans.* D67.81¼.

(32.) If 2 cords of wood cost D7.50, how much must I pay for 16½ cords?   *Ans.* D61.87½.

(33.) If 5 acres will produce 78 bushels, what quantity will 14½ acres produce?

(34.) If 1 cwt. of sugar cost D11.12½, what will 7 cwt., 2 qrs., 8 lbs., cost?

(35.) If the annual income of a clerk is D1292, and he expends D2.50 per week, how much will he save at the end of the year?   *Ans.* D1152.

(36.) Bought 11 cheeses, each weighing 35 lbs., 8 oz., at 9 cts. per pound; required the cost.   *Ans.* D35.14½.

(37.) What is the value of 2 hhds. of sugar, each weighing 6 cwt., 2 qrs., 12 lbs., at D9.86 per cwt.?   *Ans.* D130.29.2.+

(38.) At 8½ cts. per pound, what will 4 cwt., 2 qrs., 12 lbs. of rice come to?

(39.) Purchased a hogshead of wine for D75; lost 10 galls.: at how much must I sell the remainder per gallon to get the first cost?   *Ans.* D1.41.5.+

(40.) If a young man spend 9 cts. for beer, and 6 cts. for cigars, daily for 1 year, how much money will he throw away?—(365 days.)   *Ans.* D54.75.

(41.) At D1.37½ per week for board, how many weeks can you board for D71.50?   *Ans.* 52.

(42.) At 3 cts. per gill, what will be the cost of a barrel of port wine?   *Ans.* D30.24.

(43.) When butter is 18.75 cts. per pound, how many pounds can you have for D9.37½ cts.?   *Ans.* 50.

(44.) If a ship sail 175 miles in 12 hours, how many hours will it require to sail 87½ miles?   *Ans.* 6.

(45.) If 15.75 yards of cloth cost D16, what will 3 yards cost?   *Ans.* D3.04.7.+

(46.) If 6 A., 2 R., 18 po., of land cost D287.50, what will 27 A., 1 R., cost?   *Ans.* D1184.78.2.+

(47.) If ⅛ of a ship is worth D2780.50, what is the whole worth?   *Ans.* D22244.00.

(48.) If $\frac{1}{5}$ of a farm is worth D789.25, what is the whole farm worth? *Ans.* D3946.25.

(49.) If $\frac{2}{3}$ of a yard cost 80 cts., what will 12$\frac{1}{2}$ yds. cost? *Ans.* D13.33$\frac{1}{3}$.

(50.) If 16.8 lbs. cost 87$\frac{1}{2}$ cts., what will 97$\frac{1}{2}$ lbs. cost? *Ans.* D5.07.8.+

(51.) If 1.5 lb. of sugar cost 12$\frac{1}{2}$ cts., what will 84$\frac{1}{2}$ lbs. cost? *Ans.* D7.02.+

(52.) If 21 men have built a wall in 1.75 days, how many men must be employed to build it in .25 of a day? *Ans.* 147.

(53.) If a pasture serve 40 horses 60 days, how many horses would eat it in 20 days? *Ans.* 120.

(54.) How many feet of window-glass in a box containing 288 panes, 7 by 9? *Ans.* 126.

(55.) If 9 A., 2 R., of land cost D784.50, what will 20 A. cost? *Ans.* D1651.57.8.+

(56.) If 95$\frac{1}{2}$ bush. of corn cost D63.50, what will 349. bush. cost? *Ans.* D232.05.7.+

(57.) If D50 will pay for 17$\frac{1}{4}$ yds. of cloth, how many yards can you buy for D80? *Ans.* 25$\frac{9}{10}$.

(58.) If a car on a railroad will run 25 miles an hour, what distance will it run in a week of 12 hours per day? *Ans.* 2100 m.

(59.) The Secretary of War receives a salary of D6000 per year; how much is this in one week? *Ans.* D115.38.4.

(60.) When a merchant compounds with his creditors for D28 on D100, how much will B. receive, to whom he owes D7845?

(61.) What cost 6 cwt., 2 qrs., 18 lbs., of sugar, at 11$\frac{1}{4}$ cts. per lb.? *Ans.* D85.79.0.

(62.) What will 21 A., 1 R., 7 po., of land cost at D46.75 per acre? *Ans.* D995.48.2.+

(63.) What is the cost of a stove weighing 2 cwt., 1 qr., 19 lbs., at D6.25 per cwt.? *Ans.* D15.12.2 nearly.

(64.) At 12$\frac{1}{2}$ cts. per pound, how many pounds of ham can you purchase for D125? *Ans.* (125×8) $(125.00.0 \div 12.5)$ 1000 lbs.

(65.) If you can cipher through a book of 72 pages in 70 days, how many days will be required to go through a book of 264 pages? *Ans.* 256$\frac{2}{3}$.

MISCELLANEOUS QUESTIONS,

IT must be obvious to every teacher that it is extremely difficult—indeed, next to impossible—for a writer on arithmetic to give in detail every minute operation necessary in the solution of the examples, in language and expression that shall be correctly understood by juvenile pupils; that much of this process necessarily devolves upon the *teacher*, whose province and duty it is to give *oral* explanations in connexion with examples on the *black-board*.

When the writer takes upon himself the task of actual *teaching* in place of *instructor*, he steps beyond the bounds of his duty, and assumes what properly belongs to the teacher, and with which the author has no *right* to interfere.

In learning the *tables*, we have usually adopted the practice of *classing* the pupils, and assigning a suitable *daily lesson;* then let them *read* the lesson aloud in class, observing to *keep time*, speaking slowly and distinctly, at first *after the teacher*, until the habit has become familiar, when they can *read* without his aid; they should be examined *individually*, to ascertain whether each pupil can *read* the *figures* or *words* correctly, and not repeat after others, for in this way some may merely learn by *rote*, without knowing a figure or word. However, it is presumed the pupil has been *taught* the figures before he is required to *read* them.

When the pupils are engaged in learning the tables, they should practise making figures on their slates, after copies set by the teacher, and be required to write down their sums with *neatness* and *accuracy*, and of a *suitable* size. Much depends upon this part in the beginning; for if the pupil is permitted to write his sums in a *careless* or *negligent* manner, like all other *bad habits*, he will most probably pursue it *through life*.

Nothing is more important than a right beginning; for a bad or incorrect habit in *reading, writing, speaking*, &c., contracted at this period of life, is not easily eradicated—this is true not only *physically* but *morally*.

When the class is sufficiently advanced in the *tables*, writing figures, &c., then let the teacher with his chalk give small questions or examples on the *black-board*, to be transcribed by the pupils on their slate. But before adding, the teacher should observe that they have all transcribed *cor-*

really; and if there exists any deficiency, be sure to have it *corrected* before *adding* or *multiplying*, as the case may be. After the operation is performed, let each pupil present his slate for examination, and if incorrect, renew the process until each pupil can perfectly comprehend the *meaning* and solution of the *example*, and *continue* this process with the beginning of every rule, and all difficulties will be easily overcome, and create a spirit of emulation in the class. Nothing requires more deliberate, undisturbed study— *"real thinking"*—than the science of numbers; the main object being *correctness*—a certainty that the solution is *right.* Therefore, on no account would we recommend *haste;* never *hurry* the pupil when engaged in the solution of his examples, nor should idleness or negligence be indulged. We would also recommend that the pupils, of a suitable age, transcribe their examples at large in a book prepared for that purpose, as a most effectual method of improvement in penmanship, spelling, and the use of language, and enhancing arithmetical knowledge, as well as improving in *book-keeping,* transcribing accounts, &c. This should be under the immediate eye and supervision of the teacher, and care be taken that the work be in a neat, clear, legible hand, without *spot or blemish.* By following the above directions, the teacher will be amply repaid for all his time and patience, in the *rapid* improvement of his pupils. This is the only way.

———

(The following pages will embrace a *practical* review of all the preceding rules.)

(1.) What is the sum of 98709, 60497, 8959478 ?
Ans. 9118684.

(2.) What is the sum of 7894865, 7984596, 8004291 ?
Ans. 23883752.

(3.) In 1840, Maine had 501,793 inhabitants ; New Hampshire 284,574; Massachusetts 737,696; Connecticut 309,978; Rhode Island 108,830; Vermont 291,948: the above states comprise the *New England states,* and the inhabitants are familiarly called " *Yankees;*" required the number of *Yankees* in the above-named states.   *Ans.* 2234819.

(4.) In 1840, the 4 following states contained the greatest number of inhabitants, namely: New York, 2,428,921 ; Pennsylvania, 1,724,033; Ohio, 1,519,467; Virginia, 1,239,797: required the number of inhabitants in the above-named states.   *Ans.* 6,912,218.

(5.) In 1845, the following quantities of grain were produced in the United States, namely: wheat, 108,548,000; barley, 5,160,000; oats, 163,208,000; rye, 27,175,000; corn, 417,899,000: required the whole amount.   *Ans.* 721,990,000.

(6.) According to Mr. M'Gregory, the population of the world is 812,553,713. According to Mr. Bell, this vast multitude is thus divided: whites, 440,000,000; copper-colored, 150,000,000; mulattoes, 230,000,000; blacks, 120,000,000: required the number according to the division of Mr. Bell.   *Ans.* 940,000,000.

(7.) In 1800, the state of New York had 586,756 inhabitants; Pennsylvania, 602,365. In 1840, New York had 2,828,921 inhabitants; Pennsylvania, 1,724,033: required the increase of population from 1800 to 1840 in the two states.   *Ans.* 3,363,833.

(8.) In 1800 the population of the United States was 5,305,940; and in 1840 it was seventeen millions, sixty-eight thousand, six hundred and sixty-six: required the increase of population in 40 years?   *Ans.* 11,762,726.

(9.) What is the difference between 9,889,000 and four thousand seven hundred and ninety-six millions, nine hundred and ninety-six thousand, five hundred and eighty?   *Ans.* 4,787,107,580.

(10.) What is the difference between eleven millions, four hundred and ninety thousand, nine hundred and fifty-one, and twelve hundred thousand, six hundred and twenty-five?

(11.) What is the difference between 950 multiplied by 25, and 475 multiplied by 50?

(12.) If 1 sheep cost D1.75, how much must you pay for 569 sheep?   *Ans.* D995.75.

(13.) How many times will a wheel turn in 1284 miles, supposing it to turn 289 times in 1 mile?   *Ans.* 371076.

(14.) To square a number is to multiply it by itself, as 5×5=25; then what is the square of 9847? *Ans.* 96963409.

(15.) To cube a number is to multiply the number by itself twice, as 3×3=9×3=27; then what is the cube of 644?   *Ans.* 267089984.

(16.) Distribute 7984 dollars, prize-money, among 74 sailors, so that each man shall have his share.
                    Ans. D107.89.1.+

(17.) If you can raise 290 bushels of potatoes from one acre of land; how many bushels would a farm of 270 acres produce? and how much would they be worth at ¾ of a dollar per bushel? *Ans.* D58725.00.

(18.) Wethersfield, in Connecticut, is famous for raising onions; a dealer in the article purchased 8795 bushels for the Philadelphia market; he paid 33 cents per bushel, and and sold them for 55 cts. per bushel, and it cost him 50 dollars to take them to market: did he make or lose by the bargain, and how much? *Ans.* D1984.90 gained.

(19.) There is a peach-orchard in Delaware of 40 acres, 4 trees on every square rod; now a fair yield would be a bushel and a half to each tree, and the average price ⅞ of a dollar per bushel: required the number of trees, the number of bushels of peaches, and value? *Ans.* D33600 value.

(20.) In the vicinity of Cincinnati (Ohio) immense fields of strawberries are cultivated for the market; one man on the Licking river opposite the city, in Kentucky, has a field of 80 acres, and an acre will produce 500 baskets, worth 15 cents per basket: required the number of baskets of berries, and sum received. *Ans.* sum D6000.

(21.) Divide 28947845784 by 432=(6×8×9). *Ans.* 67008902+122.

(22.) Divide 749805764217 by 756 (12×7×9). *Ans.* 991806566+321.

(23.) Multiply 607958479 by 84, and divide their product by 9. *Ans.* 5674279137+3

(24.) Multiply 490789645 by 704, and divide their product by 15. *Ans.* 23034394005⅓.

(25.) 120 men have to pay for a public edifice that cost thirty-five thousand, seven hundred dollars; required the amount each man has to pay. *Ans.* D297.50.

(26.) How many times is 64 contained in 95847986128? *Ans.* 1497624783⅓.

(27.) How many times is 144 contained in 876596745188? *Ans.* 6087477397+28.

(28.) The distance from London, in England, to New York, is 3570 miles; if a steamship can make the passage in 14 days, how many miles will she sail in a day? *Ans.* 255.

(29.) Reduce 1 T., 6 cwt., 2 qrs., 12 lbs., to drams (avoirdupois). *Ans.* 762880.

(30.) Reduce 14 cwt., 2 qrs., 18 lbs., 9 oz., 8 dr., to drams. *Ans.* 420504.

(31.) Reduce 4 qrs., 27 lbs., 12 oz., 11 dr., to drams.
                                        Ans. 35787.
(32.) Reduce 67896429 drams to pounds.
                    Ans. 265220 lbs., 6 oz., 13 dr.
(33.) Reduce 789645798 ounces to cwt.
                Ans. 440650 cwt., 2 qrs., 6 lbs., 6 oz.
(34.) Reduce 649542 pounds to tons.
                Ans. 289 T., 19 cwt., 1 qr., 26 lbs.
(35.) Reduce 27 lbs., 9 oz., 14 dwt., 19 grs. (troy), to
grains.   Ans. 160195.
(36.) Reduce 9 oz., 16 dwt., 14 grs., to grains.   Ans. 4718.
(37.) Reduce 67098764 grains to pounds.
                Ans. 11649 lbs., 1 oz., 1 dwt., 20 grs.
(38.) Reduce 645896 ounces to pounds.
                        Ans. 53824 lbs., 8 oz.
(39.) Reduce 7054670 pennyweights to pounds.
                    Ans. 29394 lbs., 3 oz., 10 dwt.
(40.) Reduce 29 lbs., 6 drs., 2 scr., 18 grs. (apoth.), to
grains.   Ans. 167458.
(41.) Reduce 18 lbs., 4 drs., 1 scr., to scruples.
                                        Ans. 5197.
(42.) Reduce 7 drs., 2 scr., 16 grs., to grains.   Ans. 476.
(43.) Reduce 2040900 grains to drams.   Ans. 34015.
(44.) Reduce 179485 scruples to pounds.
                Ans. 623 lbs., 2 oz., 4 drs., 1 scr.
(45.) Reduce 947 yds., 1 qr., 2 na., to nails.   Ans. 15158.
(46.) Reduce 10094 yds., 3 qrs., 3 na., to nails.
                                        Ans. 161519.
(47.) Reduce 67078784 nails to quarters.
                                        Ans. 16769696.
(48.) Reduce 879084568 quarters to yards.
                                        Ans. 219771142.
(49.) Reduce 6 hhds., 42 galls., 2 qts., to quarts.   Ans. 1682.
(50.) Reduce 94 galls., 2 qts., 1 pt., to pints.   Ans. 757.
(51.) Reduce 47 hhds., 16 galls., 2 qts., 1 pt., 2 gi., to gills.
                                        Ans. 95350.
(52.) Reduce 4076458 pints to gallons.
                                Ans. 509557 + 1 qt.
(53.) Reduce 7890484 quarts to gallons and hogsheads.
                        Ans. 31311 hhds., 28 galls.
(54.) Reduce 60984794 gills to gallons.
                Ans. 1905774 galls., 3 qts., 2 gi.
(55.) Reduce 9947898 pints to hogsheads.
                Ans. 19539 hhds., 30 galls., 1 qt.

(56.) Reduce 47 bush., 2 pks., 2 qts., to quarts.
Ans. 1522.

(57.) Reduce 209 bush., 3 pks., 6 qts., 1 pt., to pints.
Ans. 13437.

(58.) Reduce 640984 pints to pecks, to bushels.
Ans. 40061 pks., 4 qts.; 10015 bush., 1 pk.

(59.) Reduce 309847 quarts to bushels.
Ans. 9682 bush., 2 pks., 7 qts.

(60.) Reduce 498489 pints to pecks.
Ans. 3115 pks., 4 qts., 1 pt.

(61.) Reduce 3 furlongs to inches and barley-corns.
Ans. 71280 bar.-c.

(62.) Reduce 2 miles, 4 furlongs, to inches.
Ans. 158400.

(63.) Reduce 97 mi., 4 fur., 14 po., to poles.
Ans. 31214.

(64.) Reduce 18 mi., 1 fur., 18 po., 4 yds., to yards.
Ans. 31999+4.

(65.) Reduce 150 fur., 19 po., 5 yds., 5 ft., 10 in., to inches.

(66.) Reduce 107948564 inches to furlongs.
(67.) Reduce 7904856 feet to poles.
(68.) Reduce 1098476 poles to miles.
(69.) Reduce 17694897 inches to miles.
(70.) Reduce 94876 acres to roods and poles.
Ans. 379504 R., 15180160 po.

(71.) Reduce 849460 acres, 2 poles, to poles.
Ans. 135913680.

(72.) Reduce 940 A., 3 R., 30 po., to poles.
Ans. 150550.

(73.) Reduce 74 A., 1 R., 20 po., 8 yds., to yards.
Ans. 359983.

(74.) Reduce 1468940 poles to acres.
Ans. 9180 A., 3 R., 20 po.

(75.) Reduce 150896 poles to roods and acres.
Ans. 3772 R., 16 po.; 943 A.

(76.) Reduce 140 feet of timber to inches. *Ans.* 241920.
(77.) Reduce 1920 cubic feet to cords. *Ans.* 15 C.
(78.) Reduce 3058560 cubic inches of round timber to tons. *Ans.* 44 T., 10 ft.

(79.) Reduce 14 cords to feet and inches.
Ans. 1792 ft., 3096576 in.

(80.) Reduce 10800 minutes to signs. *Ans.* 6.
(81.) Reduce 4 sig., 14°, 20′, to minutes. *Ans.* 8060.

(82.) Reduce 30°, 40′, 18″, to seconds.   Ans. 74418.

(83.) Reduce 978496860 seconds to days.
          Ans. 11325 da., 4 h., 41 m.

(84.) Reduce 41 days to hours, minutes, seconds.
          Ans. 984 h., 59040 m., 3542400 sec.

(85.) Reduce 1 yr., 7 mo., 8 da., 14 h., to hours.   Ans. 13886.

(86.) Reduce 25 yrs., 9 mo., 16 h., 18 m., to minutes.
          Ans. 445938.

(87.) Reduce 11 mo., 29 h., 14 m., 40 sec., to seconds.
          Ans. 28592080 sec.

(88.) Reduce 9 w., 8 da., 14 h., 18 m., 24 sec., to seconds.
          Ans. 6193104.

(89.) Reduce 4196784 minutes to weeks.
          Ans. 416 w., 2 da., 10 h., 24 m.

(90.) Reduce 97846849 seconds to years.
          Ans. 3 yrs., 4 mo., 1 w., 4 da., 11 h., 40 m., 49 sec.

(91.) Reduce 4180658 hours to years.
          Ans. 477 yrs., 89 da., 2 h.

(92.) Reduce 84978 hours to weeks.
          Ans. 505 w., 5 da., 18 h.

(93.) There are 4 pieces of cloth; the 1st contains 18 yds.,
1 qr., 2 na.; 2d, 16 yds., 0 qr., 3 na.; 3d, 24 yds., 2 qrs., 0
na.; 4th, 28 yds., 1 qr., 2 na.: required the number of yards
in the 4 pieces.   Ans. 87 yds., 1 qr., 3 na.

(94.) In 2 pieces of cloth, one of 28 yds., 2 qrs., the other
29.5 yds., how many yards? and how much will it come to
at D2.50 per yard?   Ans. 58 yds.; D145.00.

(95.) In one field of 9 A., 2 R., there was produced 150
bush., 2 pks., of rye; in another of 6 A., 14 po., 115 bush.,
3 pks., of wheat; in another of 8 A., 3 R., 400 bush., 1 pk.,
of corn: required the number of acres in the three fields,
and the quantity of grain raised.
          Ans. 24 A., 1 R., 14 po.; 666 bush., 2 pks.

(96.) I purchased 3 lots of wood; the 1st contained 8 C.,
28 ft.; 2d, 9 C., 40 ft.; 3d, 12 C., 60 ft.: required the quantity purchased.

(97.) A grocer purchased 4 hhds. of sugar; the 1st contained 8 cwt., 1 qr., 18 lbs.; 2d, 9 cwt., 3 qrs., 24 lbs.; 3d,
11 cwt., 1 qr., 6 lbs.; 4th, 8 cwt., 2 qrs., 25 lbs.: required
the number of pounds, and value at 8 cts. per pound.
          Ans. 4301 lbs.; value D344.08.

(98.) Add 8 sig., 4°, 4′, 28″; 9 sig., 18°, 15′, 20″; 7 sig.,
18°, 58′, 54″; 7 sig., 12°, 50′, 45″, together.
          Ans. 32 sig., 24°, 9′, 27″.

(99.) In a family of 4 children, the age of the eldest is 8 yrs., 9 mo., 7 da., 24 h.; the 2d, 6 yrs., 7 mo., 4 da., 20 h.; 3d, 4 yrs., 6 mo., 2 da., 18 h.; 4th, 2 yrs., 5 mo., 4 da., 6 h.: required the sum of their united ages.

(100.) 1 T., 4 cwt., 2 qrs., 18 lbs., 12 oz.; — 5 cwt., 3 qrs., 25 lbs., 12 oz.

(101.) 250 T., 12 cwt., 1 qr., 18 lbs.; — 156 T., 18 cwt, 3 qrs., 16 lbs.

(102.) 19 cwt., 1 qr., 18 lbs., 11 oz., 4 drs.; — 5 cwt., 3 qrs., 12 lbs., 14 oz., 8 drs.

(103.) General Hull surrendered his army at Detroit (Michigan) August 16, 1812; and General Taylor gained the victory of Palo Alto, May 8, 1846: what length of time elapsed from the surrender of Hull to the victory of Taylor? *Ans.* 33 yrs., 8 mo., 23 da.

(104.) The battle of Bennington was fought on the 16th of August, 1777; and General Cornwallis surrendered his army to the American forces at Yorktown (Virginia), October 19, 1781; what time elapsed between the victory at Bennington, and the surrender of Cornwallis.

(105.) How many years was it from the close of the revolution, April 19, 1782, to the treaty of peace between the United States and Great Britain, signed at Ghent, in Belgium, December 24, 1814. *Ans.* 32 yrs., 7 mo., 5 da.

(106.) Two steamships sail from New York at the same time for Liverpool, and steer the same course; after being 4 days out, one had made 250 L., 2 m., 6 fur., the other 234 L., 1 m., 7 fur.: required the difference of distance sailed.

(107.) It is said that some birds fly at the rate of 150 miles in an hour; how long would it take to fly around the world at that rate, allowing the distance to be 25,000 miles? *Ans.* 166 h., 40 m.

(108.) Multiply 12 C., 17 ft., 110 in., by 48. *Ans.* 582 C., 51 ft., 96 in.

(109.) Multiply 18 T., 1 cwt., 2 qrs., 12 lbs., by 25. *Ans.* 452 T., 0 cwt., 0 qr., 20 lbs.

(110.) Multiply 40 bush., 2 pks., 4 qts., 1 pt., by 36. *Ans.* 1463 bush., 0 pks., 2 qts., 0 pt.

(111.) Multiply 4 yrs., 6 mo., 1 w., 4 da., 11 h., by 96. *Ans.* 435 yrs., 3 mo., 1 w., 1 da., 0 h.

(112.) Multiply 25 A., 1 R., 14 po., by 54. *Ans.* 1368 A., 0 R., 36 po.

(113.) Divide 78 lbs., 7 oz., 12 drs., by 28. *Ans* 2 lbs., 12 oz., 13 drs.+16.

(114.) Divide 872 cwt., 1 qr., 18 lbs., 14 oz., by 96.

Ans. 9 cwt., 0 qr., 9 lbs., 13 oz.+14.

(115.) Divide 1515 A., 1 R., 18 po., by 9.

Ans. 168 A., 1 R., 19 po.+7.

(116.) Divide 341 mo., 3 w., 6 da., 18 h., by 6.

Ans. 56 mo., 3 w., 6 da., 23 h.

(117.) Divide 221 m., 4 fur., 30 po., by 15.

Ans. 14 m., 6 fur., 7 po.+5.

(118.) When rye is selling at 75 cts. per bushel, how many bushels can you have for D150?　*Ans.* 200 bush.

(119.) A gentleman purchased 2106 A., 3 R., 10 po., of new land; he wishes to divide it equally among his 5 sons, so that each may have a good farm: how much will each receive?　　　*Ans.* 421 A., 1 R., 18 po.

Note.—It is presumed that the pupil has solved *all* the preceding examples, and answered the questions in the *reviews.* If this is the case, he is prepared to use the large *" Calculator,"* which, if faithfully studied, will make the *competent* and *finished* ARITHMETICIAN.

THE END.

RECOMMENDATIONS
TO THE "COLUMBIAN CALCULATOR."
(TICKNOR'S ARITHMETIC.)

POTTSVILLE—PUBLISHED BY BENJAMIN BANNAN;

PHILADELPHIA—DANIELS & SMITH; NEW YORK—J. S. REDFIELD, CLINTON HALL; AND FOR SALE BY ALL THE BOOKSELLERS.

FROM several hundred letters commendatory of the "Columbian Calculator," from gentlemen of learning and respectability, and residents of different sections of the Union, the following have been selected, which will be sufficient to satisfy any person of the value and merits of the work, so far as recommendations can be relied on. As far as this work is known, it has received universal approbation, and is considered by teachers as the beginning of a new era in this department of science, and a desire has been expressed that the old system of confining the pupil for years in the process of reducing *pounds to farthings*, and *farthings to pounds*, should be discontinued. It is also the opinion of the best informed teachers, that the use of those books composed chiefly of a foreign currency should be prohibited in our schools, as their use is believed to be a waste of time and money, without the least benefit or advantage to any one. Those who have examined the work, and many who are using it in their schools, can speak for themselves.

From Dr. Ruschenberger, M. D., Surgeon U. S. Navy, Brooklyn, N. Y.
U. S. Naval Hospital, New York, May 9, 1845.

A. TICKNOR, ESQ.—*Dear Sir:* I have examined with some attention the "Columbian Calculator," prepared by you for the use of schools. It gives me great pleasure to believe your system of decimal arithmetic is better adapted to the daily business wants of the people of the United States than any work on arithmetic with which I am acquainted. The examples and illustrations of the several rules are well devised and American in their character. I should be glad to know that your book is extensively used in our primary schools. Respectfully and truly your ob't serv't.

W. S. W. RUSCHENBERGER.

High School, Newburgh, New York, 1846.

Having examined Mr. Ticknor's Arithmetic, with considerable care, I have come to the conclusion that it is well calculated to impart a full and clear understanding of figures, as applicable to the business transactions of the country. Its particular superiority over other arithmetics of the day consists in its lucid illustrations and correct application to business: the currency of the United States. C. M. SMITH, *Principal.*

I cheerfully concur in the sentiments expressed by Mr. Smith in the above. JACOB C. TOOKER, *Classical Teacher, Newburgh.*
M. STEVENSON, M. D., *Principal, Public School.*

Franklin, New York, 1846.

After having examined a copy of the "Columbian Calculator," I am happy to record my opinion in its favour, as a work much needed in our district schools. I have long desired to see a work that would release us from a system of arithmetic that is never brought into practical use. From its plain, simple and practical examples, I think it is just the thing we want.

Yours, &c., GEO. GREEN, *Teacher of the Village School.*

Dear Sir: Be pleased to accept my thanks for the favour you have done me in presenting me with a copy of the "Columbian Calculator." I have examined it, and consider it well adapted to the use of our district schools and academies, and recommend it as such to all teachers who wish to improve their pupils in *practical arithmetic.* Yours, truly, J. R. KNAUSS,
Principal of Bethlehem District School, Pa., 1846.

Brooklyn, New York, 1845.
I have examined a treatise on arithmetic written by Mr. A. Ticknor, and do not hesitate to say that I *like* his arrangement, and believe it well calculated to aid the teacher in imparting instruction in this branch of science.
HENRY DEAN, *Principal of Public School No. 7.*

I have examined the treatise on arithmetic above referred to, and believe it well adapted for common school instruction. S. C. BARNES,
Principal of Public School No. 4, Brooklyn.

I take pleasure in concurring in the above recommendations.
JONA. HUNTINGTON,
Principal of Public School No. 8, Brooklyn.

I have examined Mr. Ticknor's work, entitled the "Columbian Calculator," and am highly pleased with the manner of its execution, and intend to introduce it into my school as soon as practicable. M. S. PIXLEY,
Principal of High School, Gothic Hall, Brooklyn.

I have examined Mr. Ticknor's arithmetic, and think it well calculated for common schools and academies, and recommend it as a work well worthy the patronage of the public. FREDERICK SEDGWICK, A. M.
Principal of the Salisbury Academy, Connecticut.
J. J. NORTON, H. PRATT, WM. WRIGHT,
Teachers, Salisbury, Connecticut.

I have examined Mr. Ticknor's work on arithmetic entitled the "Columbian Calculator," and consider it admirably calculated for common school instruction. As soon as practicable I will introduce it into my school.
H. DALES,
Principal, Classical School, 13th Street, Philadelphia.

Dear Sir: I have thoroughly examined your work on arithmetic, the "Columbian Calculator," and pronounce it THE BOOK, for the common schools and academies of this country. The arrangement, the number of *practical examples,* and the full, and explicit explanations of the rules, render it well calculated to impart a thorough knowledge of this most important science. I will introduce it into my school as soon as convenient.
I am, with respect, A. KIRKPATRICK,
Classical School, Easton, Pa.

The above recommendation of Mr. Ticknor's "Columbian Calculator," concurs with my views of its decided excellence as a work on practical arithmetic, for use in common schools. I shall introduce it into my school without delay.
HENRY GRIFFITHS,
Teacher Young Ladies' School, Easton, Pa.

Flemmington, New Jersey, 1845.
have been favoured with the examination of the "Columbian Calculater," a treatise on elementary arithmetic by Almon Ticknor. The plan and arrangement of the work appears to me to be well adapted to the pur-

poses for which it is designed by the author. The rules are comprehensive and the examples are arranged in a natural and progressive order, eminently calculated to facilitate the advancement of pupils in that important branch of education, and what farther recommends it to public favour, it is a work entirely practical. It ranks in my estimation among the best books of the kind now in use in our district schools.   JOHN CHAPMAN,
Principal of the Flemmington Academy.
A. B. CHAMBERLAIN, A. WILSON, *Teachers.*

So far as my examination of the "Columbian Calculator" has gone, I am well pleased with the book, I think the author's views regarding the most efficient mode of teaching arithmetic altogether correct, and would like to encourage the greater use in schools of such books as that of Mr. Ticknor's.
S. F. BURNS, *Classical Institute, George st., Phila.*
Also from—B. A. LEWIS, *Classical Teacher, S. 2d st.*
D. CALDWELL, *Classical Teacher, Schuylkill 7th st.*
A. B. IVANS, *Teacher Friends' School, Salem, N. J.*

The "Columbian Calculator," by Mr. Almon Ticknor, I have examined, and must say that I am very much pleased with its arrangement, and think its introduction into our schools would be a great and public benefit.
FRANCIS WINDSER,
Principal, Public School No. 1, Hudson, N. Y.
I concur in the above recommendation.   T. H. STOUT,
Principal Public School No. 3, Hudson.

Reading, Pa., 1845.
I have to some extent examined the "Columbian Calculator," by Mr. Ticknor, and without hesitation express my belief, that it is a most admirable work, in the hands of a proper teacher, to give the young scholar a proper idea of the science of arithmetic. There is ample room yet here for some minds that have closely studied the operations of the mental faculties of the youthful scholar, to supply a work on arithmetic that will ease the labour of the teacher, and aid the pupil in his studies. In the "Columbian Calculator, this desideratum has been supplied, and the work receives my most hearty approbation.   B. F. STEM, A. M.
Principal of a Select Classical School.

New York, 1845.
I have examined a work on arithmetic entitled the "Columbian Calculator," by Mr. Ticknor, am well pleased with the arrangement and the general mode of elucidating the principles of numbers. It is a work which I have no hesitation in recommending to the friends of education every where.   HENRY SWORDS,
Principal of English Academy, No. 38 Sixth Avenue, N. Y.
I most cheerfully concur in the above,
MYRON BEARDSLEY, 147 *Waverly Place, N. Y.*

I have heretofore, in more than one instance perhaps, rather incautiously recommended and adapted school-books, which on mature and deliberate experiment have utterly failed to meet the exigency of the occasion. But from the examination I have made of Mr. Ticknor's "Columbian Calculator, or Practical System of Decimal Arithmetic," I feel an assurance that in *this case,* at least, experiment will *fully justify* my approbation. In proof of the sincerity of my conviction, I shall introduce it on the very first opportunity.   J. A. GASTON,
Principal, Somerville Academy, N. J.
Also from—
GILBERT PILLSBURY, *Principal, Classical School, Somerville.*

I have examined your "Columbian Calculator," with some care, and unhesitatingly give it the preference of any work of the kind now in use, with which I am acquainted. Sufficiently concise without being abstruse; and sufficiently perspicuous without redundancy, it seems admirably adapted to the sphere for which it was intended: and is worthy of the attention of all who feel an interest in the prosperity of our common schools.

Respectfully yours, LEVI BAILEY,
Town Supt. Lexington, N. Y.

Sir: Agreeably to your request, I have examined the "Columbian Calculator," and feel no hesitation in saying that I am highly pleased with the construction of the work; and I do further certify, with entire cheerfulness, my opinion that the work is of the greatest value, and that, in preparing it, the author has rendered to the science of arithmetical learning an estimable service. Respectfully yours, WM. MASON,
Principal of Kutztown Boarding School, Pa.

Sheffield, Mass., 1846.
I have examined M. Ticknor's "Columbian Calculator," and have been much pleased with its general arrangement, and especially with the conciseness and simplicity of its rules. The rejection of the "*pounds, shillings, and pence,*" in a great measure, and the introduction of DECIMALS, in their proper place, is an improvement. I hope to see the labours of the author repayed by its general introduction into our schools and academies.
J. M. SHERMAN, *Principal of Sheffield Academy.*
Also from—A. ROLLINS, *Preceptor of Sheffield High School.*

Schuylkill County, Pa., 1847.
In regard to your Arithmetic, the fact is this, and it cannot be controverted: it contains so many *practical* rules and *examples,* specially adapted to the various *arts and trades* of our country, (as well as its thorough illustration of our *American currency,* to the exclusion of *pounds and shillings,)* that ARTISANS and MECHANICS will make it the "man of their counsel," in all their numerical calculations. It also possesses another advantage (not the least) above all others, which will greatly facilitate its sale and introduction—IT IS CHEAP. It must and will supersede those superficial works with which the country is too much supplied. Yours, &c.,
ARTHUR WYLIE, *Teacher.*

Also from—GEO. W. GOOD,   A. S. JAMES,
      DANIEL LYONS,   JOHN M. CULP,
      A. H. LUNG,   C. THATCHER, *Teachers.*
      DANIEL STODDARD,   F. H. SMITH, *Prof. Math.*
      A. B. LUNG,
Also recommendations received from—
B. MALONE, M. D., *Bristol, Pa.*
M. S. INGERSOLL, *Principal of West Stockbridge Academy, Mass.*
Dr. LEBARR, *Principal Female Seminary, Poughkeepsie, N. Y.*
JOHN VANVALDENBERGH, *Principal High School, Allentown, Pa.*
IRA OLMSTEAD, *Teacher and Sup't, Orange Co. N. Y.*
Rev. Dr. EDWARD COOKE, *Principal Male Seminary, Pennington, N. J.*
ALEX. JOHNSTON, *County Sup't, Orange Co. N. Y.*

Reading, Pa., 1845.
Having examined Mr. Ticknor's Arithmetic, I am fully persuaded it possesses *merits* of no ordinary kind, which ought to enlist the attention of teachers of youth on the all-important science of *Common Arithmetic.* I shall introduce it into my school without delay.
B. M. HOAE, WM. GILBERT, and JOHN KELLEY,
Principals of Public Schools.

I think Mr. Ticknor has well succeeded in comprising much in a small space, in his "Columbian Calculator." It strikes me favourably in several respects, particularly the "Reviews," and the "Appendix," where "artificers' work is arranged under different heads. The book is decidedly practical, the greatest recommendation that can be given to an arithmetic designed for our district schools and academies.

J. G. MARCHAND, A. M.
Principal of Myerstown Academy, Lebanon county, Pa.
LYMAN NUTTING,
Principal of the English Department, Myerstown Academy.
Also from—J. W. KLUGE, A. M., *Principal of Lebanon Academy, Pa.*
SAML. GREENWALT, *Princ'l Select School, Lebanon Bor. Dist.*
CYRUS S. RAMSEY, *Principal of Public School, Lebanon, Pa.*

Having examined the "Columbian Calculator," by Mr. Ticknor, which I believe to be the best system of arithmetic now in use in our district schools, I would recommend its immediate introduction into the schools of this borough. N. RANK, M. D.,
President of the Board of School Directors, Lebanon, Pa.

Bristol, Pa. 3d mo. 30, 1847.
This is to certify that, after having examined the "Columbian Calculator," by Mr. Ticknor, that I am free to say that I believe it to be one of the best works of the kind that has yet been offered to the public, and cheerfully recommend it to the attention and patronage of all those engaged in teaching the important science of Arithmetic. J. V. BUCKMAN,
Principal of the Bristol Public School.
Also from O. H. HAZARD, *Princ'l of Public School, Morrisville, Pa.*

Luzerne County, Pa., 1846.
Sir:—I have examined the "Columbian Calculator," and can say that I think it a work well adapted for use in our district schools and academies. The arrangement of the rules, and examples to illustrate those rules, is, I think, particularly favourable to the progress of learners, especially those whose opportunities are few ; and it will give them a *business knowledge* of arithmetic with comparatively little exertion on their part. Yours, &c.,
A. KETCHAM, JOHN ENGLE,
WM. PIERSON, JOHN VAN LIEU, *Teachers.*

Hackettstown, N. J., 1846.
I have examined Mr. Ticknor's Arithmetic, and consider it *second* to no work of the kind now in use in our schools. I will introduce it into my school as soon as practicable. JOHN F. HUNTER,
Principal of Hackettstown Academy.

Reading, Pa., Feb. 26, 1847.
MR. TICKNOR—*Dear Sir :* From a cursory perusal of your judiciously constructed Arithmetic, the "Columbian Calculator," I am satisfied that, by its adaptation and appropriateness, it may be as useful in acquiring a correct knowledge of this important branch of education, as the most labour-saving machine has proved to be in the business of manufacturing or agriculture, and would most cheerfully recommend it to the friends of education.
Yours, &c., S. W. STEWART,
Professor of Penmanship, Book-Keeping, &c.
Also from—
G. F. SPAYD, *Principal West Ward Grammar and Mathematical School.*
FRANCIS WASSELS, *Principal of Classical School, Womelsdorf, Pa.*
J. S. NUTTING, A. M., *Principal of Womelsdorf Union Academy.*
DAVID STEACH *Principal of the Public Schools, Womelsdorf.*

Mechanicsburg, Feb. 1841.

Dear Sir: From a cursory perusal of your concise and practical system of Arithmetic, I feel free to state that the simplicity and lucidness of the statement, and the clearness of the illustrations of them, as well as the naturalness of the order of the work, embracing all that is *valuable* and *necessary* in a system of practical Arithmetic, strongly recommend it to the American people for the adoption of it in their schools for popular education.

   Yours very sincerely,   JACOB WEAVER, M. D.,
       President of the Board of School Directors.

Also from—
   JOHN HINCLE, *Principal of the Public Schools.*

March, 1847.

 From a cursory perusal of the "Columbian Calculator," I have formed a more favourable opinion of it than of any other new Arithmetic I have seen. Its rules are expressed with accuracy and precision, the illustrations are concise and perspicuous, and the examples are sufficiently numerous for all practical purposes. I would be pleased to see it supplant many works on the same subject now in use.

   JOHN KILBOURN, A. M., *Principal of Newville Academy, Pa.*
   JAMES M. KEEHAN, *Principal of Public School, Newville.*

South Easton, Pa. 1845.

 The "Columbian Calculation," by Mr. Ticknor, is an admirable work, and one which the wants of the common schools have long required. I have no hesitation in pronouncing it decidedly superior to any work of the kind I have ever examined, and I hope to see it universally introduced into our schools.   THEODORE B. FAIRCHILD, *Classical Teacher.*

 Dear Sir: Having examined the "Columbian Calculator" with some care, I feel no hesitation in recommending it to the notice of the community generally, and especially to the directors and trustees, and teachers of our district schools and academies, as it appears to me calculated, in an eminent degree, to facilitate the study of that most useful and important branch of common school education. The introduction of a treatise on DECIMALS, in the early part of the work, and the exclusion of calculations in sterling money, are features which must address themselves favourably to all; while the copious collection of examples following each rule makes it more full and complete than any other work yet offered to the public. Hoping that your work may meet that *speedy* introduction to public favour which it so richly merits. I am, with much respect, yours, &c.

       JOSEPH MIFFLIN,
   Principal of Male High School, Shippensburg, Pa.

Carlisle, Pa., March 22, 1847.

 Dear Sir: After a careful examination of the "Columbian Calculator," I feel at liberty to say that I regard it as admirably adapted to the purpose of common school education. The rules are well expressed, and the different steps of every process so distinctly marked, that the student is in no danger of confounding them together, or mistaking their order. This is but one of the recommendations of the work. The others will become obvious upon a fair *trial.* I have no doubt that its faithful study will go very far towards laying the foundation of a solid acquaintance with mathematical science.   Very respectfully yours,   A. DEVINNEY,
     Principal of the Male High School, Carlisle, Pa.

Also from—
   P. QUIGLEY, *Teacher of Public School No. 10.*
   D. ECKLES, *Teacher of Public School.*
   A. W. LOBACH, *Teacher of Public School.*

Dear Sir: In reply to your inquiry, I can say that I have been engaged more than *ten years* in the cause of educating our country's children, and in that time have used nearly a *score* of arithmetics in my school; among which were Pike's, Daball's, Rose's, Smiley, Watts, Emerson, Davies, Smith's, &c.; but they are all seriously deficient in all those eminent respects, particularly in relation to *our currency*, as well as in other points. The arrangement of your arithmetic is excellent, and leaves very little chance of improvement in this respect. It is an AMERICAN ARITHMETIC, adapted to American *currency*, to American *Teachers* and *scholars*. The numerous and appropriate examples given under the respective rules, illustrating and explaining the *modus operandi*, &c., is a feature that places it above *all* other arithmetics in this respect. I can teach the children *more* *of arithmetic*, in *less time*, with *less labour* to *them* and *myself*, from this, than any other arithmetic. I have no doubt but that it will soon supersede *all* the old antiquated, obsolete, and "heterodox" systems now so widely circulated and generally used. My *children can readily see into* arithmetical principles in the use of this book; they seem to have an *instinctive* preference for it, over *all* others. I am delighted with the *quantity* and *quality* of DECIMAL MATTER which it comprehends. I am sure its explanation in *decimals* alone secure for it the widest use—as wide as our country extends.
    Yours truly,   J. N. TERWELLIGER, *Teacher, Asbury, N. J.*
  Also from—J. R. LOVELL, *Teacher and Sup't, Harmony, N. J.*
      WM. A. LODER, *Teacher and Sup't, Alexandria, N. J.*
      WM. A. HUFF, *Teacher.*

From the Hon. Jesse Miller, Secretary of State, and Superintendent of Public Schools in Pennsylvania.
    SECRETARY'S OFFICE, *Harrisburg, March* 19, 1847.
  My Dear Sir: From the examination which I have been able to give to the "Columbian Calculator," and the confidence I have in the recommendations of those who have examined more thoroughly, I have no hesitation in pronouncing it an excellent practical work, and admirably adapted to the use of our schools. I am, very respectfully, &c.   J. MILLER.
  Also recommendations from—
WM. S. GRAHAM, *Principal of the Harrisburg Academy.*
E. L. MOORE, *Principal North Ward Male High School.*
LEWIS H. GUISE, *Principal North Ward Male School.*
F. FRICKERSON, *Principal South Ward School.*
C. M. SCHEINER, *Principal High School, South Ward.*

From Counsellor Rawn.
        Harrisburg, March 19, 1847.
  Dear Sir: During the six years that I have served as a Director of Public Schools in this borough, a number of arithmetics, by different authors, have been submitted for my examination, but it seems to me that the "Columbian Calculator," pursuing the *decimal* system, and adhering to the currency familiarly denominated *federal money*, is eminently fitted for that PRACTICAL instruction which should be a PARAMOUNT object in all common schools to impart. Accompanied by the "Key," in cases for which that adjunct is *especially* designed and furnished with a *concise*, but sufficiently comprehensive *appendix*, containing the rules (with suitable examples under each head) for computing *bricklaying, masons, carpenters,* and *joiners, slaters* and *tilers, plastering,* and *painting,* and *glazier's* work; and winding up with an exposition of the rules of and variety of examples in Mensuration and Trigonometry; it is undeniably a work promotive of the *highest* convenience to learner and instructor, and admirably adapted for use in all the common schools of our county, where that instruction most PRACTICAL and USEFUL, and measurably indispensable in the every-day business of life, should obtain dominant sway. Yours, &c.,   CHAS. C. RAWN.

March, 1847.

Dear Sir : At your request, I have examined the "Columbian Calcu-lator," and am fully satisfied of its merits. I entirely concur in the opinion expressed by Profs. M'Cartney and Yeomans. The *practical character of* the illustrations, as well as the numerous examples, make it admirably adapted to the object intended. No recommendation is needed to insure the book a general circulation and introduction into our schools and acade-mies. Respectfully yours,   M. L. STOEVER,
Professor of History and Principal of the Preparatory
Department Pennsylvania College, Gettysburg.

Also from—W. WITHERAW, *Principal of Public School.*

___

March, 1847.

Having examined the "Columbian Calculator," I highly approve of it as a book exceedingly well adapted to the purpose for which it is designed. Mr. Ticknor, from long experience as a *practical teacher,* is qualified to pre-pare a volume of this nature. I will adopt it as a regular text-book in my own teaching.   OLIVER ST. JOHN,
Rector of the Academical Department of Lafayette College, Easton, Pa.

I concur in the above, and do not hesitate to say that a *circulation* is only necessary to become generally in use.   CHARLES F. THURSTON,
Principal of the Female Seminary Easton, Pa.

___

Mauch Chunk, June 7, 1847.

I deem it almost needless to add any thing to what has already been said commendatory of Mr. Ticknor's work, the "Columbian Calculator;" but being solicitous that the youth under my care should progress as rapidly as possible as they can thoroughly in the acquisition of knowledge, and re-garding an acquaintance with Arithmetic as a very essential part of that mental culture they are expected to receive while connected with this school, I have, after a cursory, but to some extent close examination of the merits of the work, determined to introduce it into this school, as being a system of Arithmetic vastly *superior* to any that has yet come under my observation. I state the fact of my doing so, hoping that others may be induced to "do likewise."   JOHN I. MORGAN,
Principal of High School, Carbon co., Pa.

Also from—G. W. DODSON, *Teacher.*

___

Mr. Ticknor—*Dear Sir :* The undersigned make use of your Arithmetic in our schools, and take pleasure in recommending the same to the friends of education generally, and to our public and private schools in particular, as a work fully entitled to the large patronage it has already, and which we cannot doubt it will continue to receive. Its special regard and adaptation to the CURRENCY OF OUR COUNTRY, presents *peculiar* claims on the patronage of our AMERICAN SCHOOLS above *all other works* of a similar character which have issued from our press. Such a work evidently has been long needed, and it fills a vacuum long felt in this department of useful and practical sci-ence. We hope it may receive that welcome into our schools to which its *originality* and merits eminently entitle it.
With respect and esteem, yours truly,
JAS. J. OKILL, *Principal of South Easton School, Pa.*
GEO. A. NICHOLS, *Teacher at Glendon, Pa.*

COLUMBIAN CALCULATOR.

MERITS OF THE COLUMBIAN CALCULATOR COMPARED WITH OTHER WORKS ON ARITHMETIC.

1st.—*Arrangement.*—The first and most important part in all mathematical works is *order*, "Heaven's first law," or *arrangement*, in the absence of which *disorder* and *confusion* must reign triumphant. In this work the *rule*, examples, &c., are so arranged and introduced as to lead the pupil forward by easy gradations from one *step* to another up the hill of science, so as not to discourage or cause him to falter or linger by the way, but to reach the summit in full strength and vigour, with the consolation that he has performed an honourable and meritorious deed. The simplicity and perspicuity of the questions and language, their peculiar adaptation to the youthful mind, together with their practical utility in the daily occurrences of life, are considerations of the greatest importance. Many questions are so prepared as to require several *statements*, or calculations; this is an improvement, and an excellent method to bring forth the powers of the mind, and exercise all the thinking faculties. The subject is presented to the *mind* and *eye* of the pupil in a clear and *bold relief*, stripped of all *verbosity, jargon, circumlocution, ambiguity*, indefinite expressions, *ball and wire work mechanism*, catechisms or questions and answers, endless repetitions, *references, riddles, songs, book-keeping, cards, dice, rum, tobacco, mesmerism, hocus pocus, legerdemain, parts, books, chapters, sections, platoons, and divisions*, a part of which abound in other works, and have been a stumbling-block and impediment to the progress of youthful knowledge quite too long, and is the chief reason why the pupils in our public schools are more deficient in this department of science than any other. It contains more valuable matter and general arithmetical knowledge; is the cheapest in proportion to its size; can be learned in half the time, and, when learned, will be retained in the mind of the pupil and applied to the business of practical life, and this cannot be said of any other work with the least regard to *veracity*. The reviews are so arranged and prepared as to render it impossible for the pupil to *go through* the work *ignorant* of its contents, if the instructor performs his duty faithfully. The unreasonable practice of some authors of introducing *questions* and *answers* ready prepared, "cut and dried," for the pupil to *answer* at the *beginning* of a *rule*, is not only inconsistent with common sense, but highly injurious; the mind is not brought into action. The *author* has prepared the *answers*, and the pupil is to *learn* and *repeat* by *rote* like a *parrot*, and with about as much intelligence. The illustrations and explanations, particularly in the more difficult and important part, are more full and perfect than can be found in other works; for instance, *Proportion, Fractions, Interest, Evolution*, &c., with the admirable examples and selections in *Artificers'* work and mensuration, most materially enhance the *value* of the work over any other.

2d.—*Currency.*—In times past when our currency was *pounds*, shillings, &c., which was the currency of this country under the *benevolent* British government, previous to the Revolution, it was important that calculations in that currency should be taught in schools; but, since the organization of our present form of government, our currency has changed with it, in *value, simplicity*, and *beauty*, in proportion as our form of government is superior to any other on earth. Therefore, instead of having our *teachers* and *pupils* employing their time in useless calculations in the currency of another country, almost to the exclusion of our own, is, to say the least, a great sacrifice, a *waste* of time and money, without the least beneficial result, which should be employed in the acquisition of some other valuable science; at least one half of their time, when devoted to the study of figures, has been employed in this way, or others of a like character, and equally foolish, such as *foreign exchange*, which is continually fluctuating and has no *fixed* value, which may be computed by *interest, proportion*, &c.; and, strange as it may appear, parents and teachers still *persist* in this injurious and *unnatural* course, and why do they do it? The answer is obvious: the books are prepared in this way, which renders them next to worthless, and the teachers have been taught from them, few of whom can use only *one*, unless it has a *key*, so that the system is handed down from father to son, like an hereditary disease, that requires the aid of the most experienced practitioner to stay or remove. Another evil and source of complaint consequent upon the deranged and unequal method of reckoning money, in passing through the several States of the Union, which frequently puts travellers to great inconvenience and trouble, is, "*pounds, shillings, bits, levies, picayunes, fips, pence*, &c., in fact every name and denomination except our own, and which have no legal existence. Why not reckon in dollars and cents at once? This is familiar and is understood by every one; even foreigners soon comprehend it better than their own, owing to its perfect simplicity. Another important feature in this work is the introduction of *decimals*, in the early part of the work, and the adaptation of those rules to our currency; this was considered of the *first* importance, and the first and only time it has been presented in this manner by any writer on the subject. If the chief design of an Arithmetic "be to learn to calculate with *precision* and accuracy in all transactions where money or property is concerned," the arrangements of the work are such as to answer the aim and object of the *author* and teacher. This is considered a decided improvement by every person in the least conversant with the subject, as the pupil will soon acquire a knowledge of decimals and his *own* currency, instead of wasting his time in striving to obtain a knowledge of a currency which can be of little or no benefit to him, then the task is comparatively easy through the following rules, as Proportion, Vulgar Fractions, &c., and, so far from considering or viewing the subject as "*dry, hard,*" and uninteresting, the opposite will be the case; he will see the propriety and utility of it, all *difficulties* will be surmounted and overcome, his *march* will be *onward*, and his success in this will encourage him to press forward in the acquisition of knowledge. It has not unfrequently been the case that scholars have become discouraged and disgusted with the science of numbers, in consequence of the stupidity of some speculating *authors*, or *verdant* teacher, neither of whom were acquainted with the science, which compels us to say, and it is beyond the power of contradiction, that no other science is so much neglected, nor so miserably taught as this, for not one *young man* in ten can calculate the *interest* on a *bond*, yet they have been to *classical schools, academies*, and some even boast of having *graduated* at some college, and claim acquaintance with the higher branches of literature. The above is a fair, just, and impartial statement of facts. Much more might be said; but "sufficient for the day is the evil thereof," and let those interested govern themselves accordingly. That this work possesses more merit, more value, more original matter, than any other work of the kind *extant*, no person in the right use of his reason (if he has any) will or dare deny, because it is better and more mathematically arranged, and better adapted to the instruction of youth in *every way*. We are warranted in saying all this, for it justifies the above assertions, and is destined with as much certainty to supersede the *worthless* systems now in use, as *truth* is to triumph over falsehood.   'H.

A great number of the public journals, periodicals, &c., in different parts of the Union, have spoken in terms of high commendation of this work. A few selections are given:

"The Columbian Calculator," by A. Ticknor, author of "The Accountant's Assistant," "Mathematical Tables," &c. &c. This is a practical and concise system of decimal arithmetic adapted to the use of schools in the United States. The copy before us is the second edition, and we hope it may reach the FORTY-SECOND. Nothing so well as treatises of this kind can aid the original intentions of the regulators of our currency, and substitute the decimal coin of the United States in lieu of the provincialisms in counting money, which prevail in different sections. Each State sets up for itself, with its "*ninepences,*" "*shillings,*" "*fips,*" "*levies,*" "*bits,*" and other out-of-the-way designations. The beauty and simplicity of decimals, their exactness and ease of computation, should lead all Americans to adopt the use of the Federal currency. This work, prepared with care, has borne the test of examination. It has been favourably pronounced upon by competent judges, and is already introduced into many schools.—*Saturday Evening Post, Philadelphia*, 1845.

We have received from Mr. Ticknor a new Arithmetic, entitled "The Columbian Calculator," of which he is the author. The arrangement of the work is admirably adapted to the comprehension of beginners, being one regular systematic advancement from the simple axioms to the most difficult problems. The use of pounds, shillings, and pence, have been entirely thrown aside—at once relieving beginners from one of the *stumbling-blocks* in the path of their arithmetical advancement. In fact, the whole work has been prepared with great care and precision, and is well worthy the attention of all persons engaged in the honourable and useful avocation of instructing the rising generation.—*Doylestown Democrat.*

Having examined a number of Arithmetics now in use, and knowing that there is a great deficiency in the system of "computing by numbers," I feel it a duty which I owe to myself and to the public generally, to recommend to the use of schools the "Columbian Calculator," as the most concise, explicit, and efficient method now extant. The author has spared no pains in selecting the most choice and easy rules, and giving the plainest problems for solution. His plan is so well calculated to obviate the various difficulties that the pupil meets with in other works, that the most complex questions are rendered simple and plain. Indeed, I have no doubt that the introduction of this invaluable work to the public is but the commencement of an era from which future authors will date a great revolution in the science of mathematics. It is evident, from the success which it has already met, that our colleges, academies, and district schools, will, ere long, adopt the author's system, to the entire exclusion of all others. The author, in selecting and arranging his system, seems to have laboured *pro bono publico*, and taken the greatest care to make his work acceptable.—(*Communicated*) *Warren Journal, N. J.*

Reform of the Currency.—Those who are in favour of seeing our own simple *dollars, dimes,* and *cents,* introduced as the only and uniform currency throughout the land, will be pleased to learn that a work will shortly appear, designed to aid in bringing about so important a matter. It is now in preparation by Mr. Ticknor of Easton, in this state.

From this gentleman's experience of *twenty* years in the responsible business of school teaching, we expect a valuable work. Mr. Ticknor desires to have a NA-TIONAL CURRENCY, such as is known to the *laws* of the land, to come into general use in place of the present *confused and mixed* method of calculation now so much in use in our mercantile transactions, and made an *indispensable* part of public instruction in our DISTRICT SCHOOLS AND ACADEMIES. The first step should be in a suitable system of ARITHMETIC, and if we cannot *compel the old* to conform to the new system, we can, at least, instruct the rising generation in the right way.—*Philadelphia Saturday American,* 1845.

Notices similar to those just quoted have been inserted in the following papers:

Sommerville Whig, N. J.	United States Gazette, Phila., Pa.
Princeton papers, do.	Saturday Courier, do. do.
Newark do. do.	Alexander's Messenger, do. do.
Trenton do. do.	Inquirer, do. do
Honesdale Democrat, Pa.	Public Ledger, do. do
Honesdale Herald, do.	Spirit of the Times, do. do.
Easton Sentinel, do.	Pennsylvanian; do. do.
Whig and Journal, do.	Wilmington paper, Del.
Journal of Commerce, N. Y.	Allentown paper, Pa.
New York Sun, do.	Highland paper, Newburg, N. Y.
New York Tribune, do.	Hudson papers, N. Y.
New York Observer, do.	Catskill do. do.
Chambersburg papers, Pa.	Gettysburg papers, Pa.
Reading do. do.	Baltimore papers, Md.
Harrisburg do. do.	Washington papers, D. C.
Lebanon do. do.	And many others.

NOTE.—The author of the "Calculator" would most *earnestly* and *sincerely* request all those interested to examine the preceding recommendations candidly and delibe-rately, without *prejudice,* and then decide; for one of two things is *self-evident*—that there is a sad deficiency in other works of the kind in regard to arrangement, quan-tity and quality of matter, currency, &c., sufficient to *prohibit* their use, unless their authors will put them in a better form and adapt them to the *American currency,* in-stead of the *English,* or those gentlemen who have spoken in such *expressive* and *decided* terms in regard to this work are *incompetent* to judge of its *merits,* and have written to *flatter* the author. There is no *medium,* no neutral ground; the author of this work has taken a *new* and independent course, *different* from *all* others in the general arrangement, *matter* and *quantity, currency, exercises,* &c. &c., equally regardless of *smiles* or *frowns,* so long as *truth* and the *public good* have been his sole aim and object. For any one to assert that "this work is just like others," or "that others contain the *same,*" is an *untruth,* and only made by those who wish to conceal their own *ignorance,* or is the result of *prejudice,* for it was the aim of the author to make it as *different* from others as the nature of the science would admit. That this is the opinion of the writers of the preceding articles, may be plainly seen, who, with few exceptions, were entire strangers to the author, consequently under no ob-ligation to confer *special favours* on him—they were given freely and willingly; aside from this, they are gentlemen of *learning, talents,* and of the first *respectability* in so-ciety, above *bribery* or *sycophancy,* and could have no possible *interest* but the advance-ment of *science* and the *public good.* Hence it follows that, if their assertions are *correct,* this volume should become the *standard* text-book of the UNION; or that this work should be *prohibited,* and *all* the others declared *perfect,* and used in all the schools. So far as this work has become known, *nine-tenths* of the most *compe-tent* teachers have given it the preference *over all others.* What does this prove?

THE

YOUTH'S COLUMBIAN CALCULATOR,

BEING

AN INTRODUCTORY COURSE IN ARITHMETIC FOR BE-
GINNERS, ADAPTED TO THE CURRENCY AND PRAC-
TICAL BUSINESS OF THE AMERICAN REPUBLIC,
FOR THE USE OF THE DISTRICT SCHOOLS,

BY ALMON TICKNOR,

AUTHOR OF THE "COLUMBIAN CALCULATOR," ETC.

THIS small volume will comprise about 84 pages, and 800 examples for solution on the slate; it will embrace the fundamental rules, Compound Rules, Simple and Compound Reduction, Single Rule of Three, or Proportion, and Simple Interest. Teachers who have examined this work in manuscript, are of opinion that it is just what is very much wanted at this time in our District Schools as a Primary Arithmetic for those commencing the study of numbers, for the reasons that those Primary Books now in use are either too *juvenile* or too far in advance of the *pupil;* in fact that there is no *suitable* Primary treatise on Arithmetic now before the public. It is also believed that this volume will contain a sufficient amount of PRACTICAL ARITHMETIC, as will commonly occur in the transaction of ordinary business —more particularly in the *Female Department* of our District Schools, many of whom seldom learn the use of numbers as far as *Reduction* or *Proportion,* and as this work is intended, in part, for this class of pupils, great care and labour have been bestowed with a view to render every part perfectly plain and easy of comprehension by the pupil. The calculations are in "*our currency,*" with the use of a few fractions, sufficient for general use, as a knowledge of *fractions* can be acquired from the *larger volume.*

The Primary Arithmetic will be published in a few weeks in Philadelphia.

The new Key to the Columbian Calculator, which will be *stereotyped* and *published* the ensuing winter, will embrace *several hundred* examples in Arithmetic and Mensuration, and other valuable matter for the use of the teacher. The *examples* will be given in *full,* with *Notes, Explanations, Illustrations, Demonstrations,* &c. &c.

The three volumes will contain about 3500 *original* questions for solution—a greater amount of *Arithmetical science* than has ever been published in the same space in this or any other country—a work that is *destined* to become the STANDARD TEXT-BOOK OF THE UNION.

J. S. Redfield, Bookseller and Publisher, Clinton Hall, New York, is Agent for the Author, who will keep the above books for sale.

P. S.—A Key to the Primary Arithmetic will be *bound* with the Key to the Columbian Calculator, or *separately.*

12

CPSIA information can be obtained
at www.ICGtesting.com
Printed in the USA
BVHW040957071218
534846BV00028B/250/P